SCIE[illegible]
MATTERS

Genetic Engineering

prepared for the Course Team
by Norman Cohen and Hilary
MacQueen

Science: a second level course

The S280 Course Team

Pam Berry (Text Processing)
Norman Cohen (Author)
Angela Colling (Author)
Michael Gillman (Author)
John Greenwood (Librarian)
Barbara Hodgson (Reader)
David Johnson (Author)
Carol Johnstone (Course Secretary)
Hilary MacQueen (Author)
Isla McTaggart (Course Manager)
Diane Mole (Designer)
Joanna Munnelly (Editor)
Pat Murphy (Author)
Ian Nuttall (Editor)
Pam Owen (Graphic Artist)
Malcolm Scott (Author)
Sandy Smith (Author)
Margaret Swithenby (Editor)
Jeff Thomas (Course Team Chair and Author)
Kiki Warr (Author)
Bill Young (BBC Producer)
External Assessor: John Durant

Authorship of this book

Chapters 1–8 and the Epilogue were written by Norman Cohen and Chapters 9 and 10 by Hilary MacQueen.

The Open University, Walton Hall, Milton Keynes, MK7 6AA.

First published 1993. Reprinted 1995.

Edited, designed and typeset in the United Kingdom by the Open University.

Printed in the United Kingdom by Thanet Press Limited, Margate, Kent.

ISBN 07492 51042

This text forms part of an Open University Second Level Course. If you would like a copy of *Studying with the Open University*, please write to the Central Enquiry Service, PO Box 200, The Open University, Walton Hall, Milton Keynes, MK 7 6YZ. If you have not already enrolled on the Course and would like to buy this or other Open University material, please write to Open University Educational Enterprises Ltd, 12 Cofferidge Close, Stony Stratford, Milton Keynes, MK11 1BY, United Kingdom.

1.2

11755C/s280geneni1.2

Contents

1 Genetic engineering — media hype or real revolution?

Wonder drugs to cure cancer, epidemics caused by laboratory-created bugs, £billion industries, high-yield wheat that grows well without fertilizer, choosing the sex of your child or the colour of his/her hair, government warnings of missing the gravy train, dire warnings of uncontrollable 'monster' weeds and the physical and moral dangers of tampering with nature, vaccines against malaria and AIDS, bacteria for mining metals or removing 'non-biodegradable' pollutants...And a whole new vocabulary virtually in everyday use—enzyme, DNA, cloning, gene, biotechnology—even spawning new advertizing gimmicks (Figure 1.1).

Whatever else it is, since its inception in the early 1970s, genetic engineering has been big news, and still is, as a glance at Figures 1.2 and 1.3 will show. Few new sciences can have been subjected to such limelight. The various and varied reports, estimates and assumptions for the products of genetic engineering range from the carefully analytical to the ludicrous.

Though genetic engineering is a science that promises wide use, it is a *new* science and so just a few practical products exist as yet—a number of new drugs, for example. So much of the excitement generated by genetic engineering revolves not around results, but around *potential*; couple this with a boom or doom style of reporting and

Figure 1.1

New Genentech Issue Trades Wildly As Investors Seek Latest High-Flier

By TIM METZ

Staff Reporter of **THE WALL STREET JOURNAL**

NEW YORK—Yesterday, shortly after 10 a.m., the paperwork connected with the public sale of one million shares of Genentech Inc., at $35 each, was completed.

An historic day in the securities markets had begun.

At 10.25 a.m., the syndicate of underwriters that had sold the shares to the public disbanded. The shares began trading in the over-the-counter market—up 128% to $80 a share. Within 20 minutes, the price had peaked at $89 a share bid, turning the original $35 million of shares into $89 million.

'I have been with the firm 22 years', said an officer of Merrill Lynch, Pierce, Fenner & Smith Inc. 'I have never seen anything like this.'

For the next five hours or so, until late afternoon, when the securities markets closed down, share prices on Genentech plunged and soared; when exhausted traders for the 27 securities firms making markets in the stock finally called it a day, the price had settled to $71.25 on volume of 528 000 shares.

A Memorable Explosion

The most striking price explosion on a new stock within memory of most stockbrokers took place in the shares of a tiny company with a barely profitable history and an uncertain future. On June 30, Genentech's total assets amounted to only $14.2 million. Its first-half earnings, before a $29 000 tax-loss carryforward, were only $51 802. But Genentech does research in one of the most glamorous sciences around, genetic engineering. The company went public at a time when investors and speculators in high-technology stocks have gained millions or even billions of dollars to plunge into such new ventures as Genentech.

For weeks, demand had been building for the one million Genentech shares that underwriters, led by the firms of Blyth Eastman Paine Webber and Hambrecht & Quist, were to sell. The Merrill Lynch officer stopped counting when his firm alone got customer requests to buy three million shares. When the shares were allocated among securities firms for sale to their customers, not a single office of Merrill Lynch—the world's largest securities firm—wound up with more than 100 shares. Some 500 of Merrill Lynch's 580 offices around the world got no shares at all.

The thousands of investors and speculators who were unable to buy at the $35-a-share offering price immediately bid up prices to $80 and above. It made a memorable day for such professional traders as Robert Antolini, at Drexel Burnham Lambert Inc.'s institutional over-the-counter trading desk.

Bucks for Bugs

'Imagine that value for a company that proposes to earn a living making bugs', Mr Antolini said yesterday morning. (The 'bugs' are bacteria that are part of Genentech's experimental work.) 'It makes you stop and think.'

But Mr Antolini has little time to think for the rest of the day. 'Eighty-six-and-a-quarter bid—its gonna rally!' he shouts to no one in particular after the shares first fall and begin to rebound. 'Eighty-six-and-a-half bid! I buy 50 at 86½,' he shouts into a telephone to a trader at another firm. In the first hour of trading, he and other Drexel traders take orders to buy 19 050 shares and to sell 15 600.

Drexel's own inventory of Genentech, or its 'long' position, rises to about 6 000 shares in the next half hour, as investor selling drives the price below $80.

Mr Antolini jumps from his chair. He jerks open a desk drawer and pops several aspirin tablets into his mouth. 'God, I've got a Genentech headache!' he complains.

Figure 1.2 From *The Wall Street Journal*, 15 October, 1980. The financial benefits of genetic engineering were soon recognized. However, as the article shows, the early money to be made rested largely on the hoped-for *potential* of the techniques, and hence on the speculative value of the newly-formed genetic engineering firms, rather than on existing assets or trading profits.

inevitably some wilder predictions might emerge. But, even on a calm analysis, genetic engineering does suggest a wide range of uses, some quite remarkable.

It is impossible in a book of this length to cover all aspects of genetic engineering in any degree of detail. This is particularly true as the science itself, its applications and the debates about it are developing at a rapid rate. But at the very least we hope to

THE INDEPENDENT

No 1,424 THURSDAY 9 MAY 1991 ★★★ Published in London 40p

DNA injections turn females into males as British researchers identify gene that determines gender

Mice embryos' sex changed

By Tom Wilkie
Science Editor

IN ONE of the most important developments in genetics in the past 40 years, British scientists have located and identified the single gene that makes the difference between men and women.

The researchers have succeeded in changing the sex of laboratory mice embryos by injecting newly fertilised eggs with a fragment of DNA, the molecule which carries the blueprint for life itself. The fragment, it has been found, genetically "switches" would-be females into males.

By studying the way the sex-determining gene acts, the researchers hope to understand better what goes on as an embryo develops, and to shed light on diseases such as cancer.

In an echo of the double helix story – when at Cambridge in 1953 the scientists James Watson and Francis Crick discovered that DNA had a structure of twin strands winding around each other – the quest for the gene which triggers maleness became something of a transatlantic race between teams in Britain and the United States. The announcement in today's issue of *Nature* clinches the victory for the British researchers.

They have discovered that the genetic essence of masculinity is a tiny strand of DNA. It is known as Sry, the sex-determining section of the Y-chromosome, and the researchers isolated it from the distinctive Y-chromosome that appears in male but not in female mammals.

The gene was identified in research done by two teams of scientists, led by Dr Peter Goodfellow, at the Imperial Cancer Research Fund, and Dr Robin Lovell-Badge, at the National Institute for Medical Research in north London.

Dr Lovell-Badge and his colleagues at the institute have injected this Sry gene into fertilised mouse eggs. Although the eggs carried two X-chromosomes – the genetic signature of being female – the extra piece of the double helix dictated that, instead of developing into females as nature had intended, the mice became male.

Dr Peter Koopman, one of the researchers at the institute, said: "It's as close as we can come to the biological proof that Sry is the

PRESS ASSOCIATION

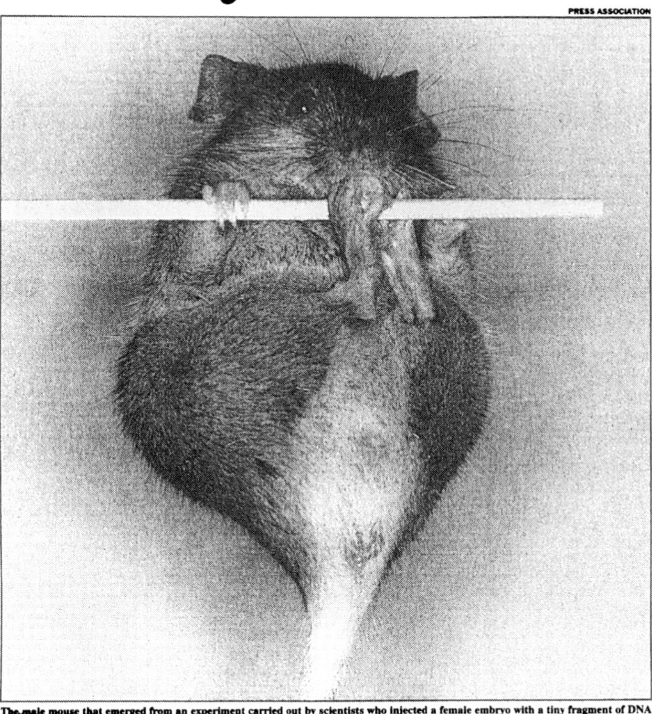

The male mouse that emerged from an experiment carried out by scientists who injected a female embryo with a tiny fragment of DNA

testis-determining factor." According to Dr Koopman, the mice experiments "show the direct link between a gene and developmental process in an embryo". He believes that the research will help scientists to understand the complex processes going on in embryos and give them a basis for understanding diseases caused by abnormal development. According to Dr Goodfellow, of the Cancer Research Fund: "It is a good investment for a cancer institute to learn more about development, and this is the most direct way for mammals. More and more genes which were described as oncogenes [cancer-causing genes] are turning out to have a role in development."

Only three out of 11 transgenic embryos developed into males; the rest went on to become normal females, but Dr Koopman believes that this may be because, in some cases, the gene integrated itself in the wrong place in the recipient DNA. The gene is very small – occupying less than one two-hundred-thousandth of the total extent of mouse DNA – and it operates for only about two days in the development of the mouse embryo before it is switched off again.

"We know that up to 11 and a half days, male and female mouse embryos look identical," Dr Koopman said. The Sry gene appears to be active in the genital region of male embryos at about 10 and a half days, but within two days – when the first signs of testis differentiation appear – it seems to be shut off.

The researchers believe that Sry contains the instructions to make a protein which activates other genes within the mouse's DNA to make all the various components required by a male.

Although the experiment was done in mice, the researchers have also isolated the sex-determining gene in humans. They flatly ruled out the possibility of their research being applied to humans to allow people to choose the sex of their own children. "We don't intend to do this on humans or anything sinister like that," Dr Koopman said.

Although the sex-reversed mice have penises and testicles and behave towards female mice as all red-blooded male mice should, they are sterile. "All mammals with two X-chromosomes are sterile, because having two Xs is not compatible with the survival of sperm," Dr Koopman said. As well as being illegal and morally objectionable, it would be totally impractical to do this on humans, he added.

GREAT NEW DEVELOPMENTS ELSPETH'

NIMR

Nationwide cuts home loan rate to 12.25%

By Patrick Hosking and Gail Counsell

A FRESH round of mortgage rate cuts looks increasingly likely after Nationwide Anglia Building Society set the ball rolling yesterday with a surprise 0.7 per cent reduction to 12.25 per cent.

The lower rate comes into effect from Saturday for new borrowers and 1 July for Nationwide's 1.2 million existing borrowers. Nationwide said that its savings rates were also likely to be cut.

Other lenders are expected to follow if base rates are reduced in the next few weeks. None was prepared to commit itself to a cut, however, without seeing a base rate fall first.

John Hutchinson, Nationwide's retail operations director, said: "We have decided not to wait for the next base rate cut." He added that he expected base rates to fall soon, but the new rates would stand in any case.

The society also announced lower rates for larger borrowers: 11.75 per cent for mortgages of more than £60,000 and 11.25 per cent for mortgages above £120,000.

The cut reduces the monthly interest payments of a borrower with a 25-year £30,000 endowment mortgage by £13.13 to £229.69. A £60,000 borrower will save £30.62.

However, the 400,000 Nationwide borrowers on mortgages which are reviewed annually will see no reduction in their repayments until January.

Jim Birrell, chief executive of the Halifax, Britain's biggest building society, said: "We intend to stay competitive. But this move has no real effect until 1 July, which is some way off. It gives us time to consider the market impact."

The Woolwich and Leeds societies accused Nationwide of a "marketing ploy" in announcing the cut seven weeks before it actually takes effect for most people. Abbey National said that it had no intention of following the Nationwide, which was taking a gamble on base rates being cut.

Mortgage rates have been falling steadily in the wake of four half-point cuts in base rates since January. Borrowers are starting to benefit from the most recent cut to 12 per cent on 12 April.

Financial markets expect an-

SUMMARY

8,500 jobs to be shed

British Coal is cutting 5,500 jobs and Rolls-Royce announced 3,000 job losses on top of 3,000 already planned .... Page 28

Rushdie attack

Salman Rushdie has attacked Islamic scholars who said he was not a Muslim .... Page 2
Letter, page 26

Pilot fined

A jumbo jet pilot found guilty of endangering his aircraft and passengers was fined £2,000 .. Page 3

Morrison guilty

Danny Morrison was found guilty of falsely imprisoning an RUC Special Branch informer .. Page 6

IT failure

A five-year government programme of information technology research and development failed in its main aim .... Page 9
Leading article, page 26

UN rebuff

The UN is rebuffing British attempts to have claims against Iraq administered in London .. Page 11

Forte switch

Trusthouse Forte may become known as Forte .... Page 28

Super league

First Division club chairmen gave cautious approval to FA plans for a super league .... Page 38

Racing cuts

The Levy Board announced £6m of cuts in expenditure on horse racing for 1992 .... Page 38

INSIDE

POP ART Images of fun and rebellion 5

FASHION 'Vogue' opens the archives 19

EDUCATION Playground power games 23

Figure 1.3 It is not often that science makes the lead headline story of a national newspaper. Though a reasonably sober, well-reported account of an advance depending on genetic engineering, perhaps the references to furry animals, a *British* 'breakthrough', cancer and sex makes this 'a story that has everything'.

present a 'sober' overview of the subject — what it actually is, what has already been achieved using it and what *reasonably* might be expected of it in the future. As you study the subject you should begin to develop your own 'critical checklist', what sorts of questions to ask when confronted with the plethora of projects, predictions and perils on offer almost daily. In such questions lies the key to sorting the wheat from the chaff, the likely from the, perhaps, more likely fantastic.

A lot of the interest generated by genetic engineering comes not so much from the science, as such, but more from its actual and promised applications. We too shall approach the subject from this angle but in so doing we shall try to identify the main elements of the science that makes these applications possible — or in some cases less likely. But, as will become abundantly evident, the use to which genetic engineering is put depends not on available science alone but also on a whole host of other factors — commercial, political, ethical and so on — and these too will be mentioned. Given the limited space available we shall concentrate largely on two areas where genetic engineering is likely to have a major impact. The first area, where some of the earliest everyday uses of genetic engineering have already come to fruition, is *medicine*. The second is *plant and animal breeding* — an area of agriculture where a good deal of current research and development in genetic engineering is focused, where a few products are emerging and where many products should enter strongly onto the market in the next ten to twenty years.

However, before tackling these two areas, we must consider some of the actual techniques of genetic engineering; this we do in Chapters 4–8. This paves the way for our two areas of application—medicine in Chapter 9 and plant and animal breeding in Chapter 10. Finally, we look very briefly towards the future and at some of the broader social and political issues surrounding genetic engineering.

Though our stated aim is to keep within the realm of fact rather than fantasy, inevitably we shall often be dealing with potential developments, hence with *predictions*. Niels Bohr, one of the founders of modern atomic theory, once remarked that 'prediction is very difficult, especially about the future'. Whether one regards this as the profound insight of a great physicist–philosopher, a subtle joke, or merely the utterance of a man not noted for his command of language, one can but agree. So before *we* predict the future, let us look briefly at what genetic engineering actually is, in principle. In doing this we shall first set it in some historical context. Although this will be done very rapidly, it will involve a fairly extensive digression into a number of traditional areas within plant and animal breeding and the use of micro-organisms. What genetic engineering offers, that traditional processes do not, will then emerge—as should the reasons for the excitement (Chapters 2 and 3).

2 Genetic engineering — why the interest?

The main purpose of this chapter is to introduce genetic engineering by showing briefly what it offers to animal and plant breeders that more traditional technologies do not. It is largely narrative in style and contains a good deal of illustrative material to make the main points. We hope that these illustrations prove of general interest but it is important that you absorb the main take-home messages. So as you read the chapter you should note down what you think the main points are and at the end of the chapter Activity 2.2 asks you to produce a summary of these.

It is not easy to cover all the aspects of genetic engineering in a single definition. But put at its simplest, **genetic engineering** is a set of techniques that enable us to transfer *specific* genes from one organism to another, *even between different species*. Furthermore, under appropriate conditions, the organisms receiving the genes can pass these on to their offspring on reproduction. In principle, this allows us to produce populations of new varieties of organisms with specific novel characteristics, those endowed by the genes transferred from other organisms.

But why should anyone want to do this?

2.1 Engineering new animals and plants

The wish to tailor other organisms to suit human needs is by no means new. Perhaps even thousands of years ago, Mexican Indians selected particularly desirable ears of maize from which to sow next year's crop (Figure 2.1). Presumably they shared a basic aim with generations of rose growers, pigeon fanciers, dog breeders and so on. They may also have shared at least one element of a basic understanding of inheritance: like begets like. And so, as all sorts of plant and animal breeders must have learnt, choosing the right parents helps produce the desired offspring. Having once obtained offspring with desirable characteristics, inbreeding — breeding within the group, between close relatives — helps ensure continuity of those characteristics generation after generation, and so maintains the pedigree.

This century, much of the folklore of plant and animal breeding has been replaced by a better-founded knowledge of the basic rules of inheritance gained from the science of **genetics**. But for plant and animal breeders the essential *procedure* is still much the same — choose parents with desired characteristics and breed offspring from them. As you will see, this is likely to change dramatically with the introduction of genetic engineering. Nevertheless a knowledge of what we can term 'traditional' genetics (i.e. genetics prior to genetic engineering) has undoubtedly contributed significantly to the success of breeders in this century (Figure 2.2).

Figure 2.1 A 3000-year-old Olmec axe-head showing a plant at the top which could be maize.

Activity 2.1

Figure 2.2 presents data on wheat yields in the UK from about 1940 to 1980. Examine this graph and then attempt (a)–(c) below.

(a) What is the percentage increase in wheat yield between 1950 and 1970?

(b) From Figure 2.2, is it possible to calculate the total annual crop of wheat in the UK in 1975, and if so what was it?

(c) The text to which Figure 2.2 pertains makes a claim for the usefulness of a knowledge of genetics in plant and animal breeding, implying that this knowledge has enabled breeders to produce more successful varieties. Does the figure in fact demonstrate that it is the use of newly-bred, better, varieties of wheat that has led to higher yields? Give concise reasons for your answer.

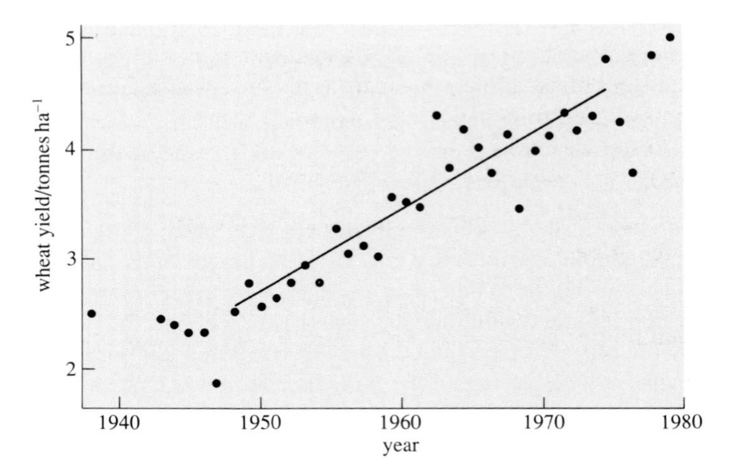

Figure 2.2 UK wheat yields over 40 years.

From Figure 2.2 alone it was not possible to say what the relative individual contributions of a variety of factors, including better varieties, might be.

▷ Figure 2.2 cannot help identify what sorts of things might have contributed to increased yields. Can you suggest some such possible things?

▶ In addition to improved varieties, over that time period it is likely that better cultivation techniques such as improved irrigation, use of fertilizers and pesticides, and a greater knowledge of soil science in general have all made contributions to increasing crop yields. (On a wider front it *may be* that greater economic incentives have also encouraged an increase in yields and in that rather different sense *contributed*.)

From Figure 2.2 alone it might be concluded that there were no new varieties or at least no contribution to increased yield from them. It is also virtually impossible to cleanly separate numerically the contribution of any one factor from any other where the factors interact in a complex way. For example, if a new variety treated with a new fertilizer and pesticide regime gives a 15% increase in yield, how does one attribute x% to the improved variety, y% to the fertilizer and z% to the pesticide if all three interact? Nevertheless, it has been *estimated* that over half the 5–10-fold increase in crop yields over the last 50 years has probably come via plant breeding.

You might also like to note that the increase in UK wheat yields has continued—the 1990 yield was about 7 tonnes ha^{-1}, some $2\frac{1}{2}$ times that in 1950 and nearly $1\frac{1}{2}$ times that in 1980.

But advances in plant and animal breeding that rest on a knowledge of traditional genetics have still failed to overcome two major barriers imposed by nature: the *randomness* of inheritance and the *species barrier*.

▷ In what senses can inheritance be said to be random?

- There is randomness in three senses.

  1 There is randomness in what parental characteristics are available to breed from. This arises because all variety in characteristics emanates *ultimately* from mutations that produce different alternative versions (i.e. alleles) of genes. And, critically, *where mutations actually occur is random.* So, for example, if there has never been a mutation in a particular gene (or, at least, you are unaware of any that exist), then there are no different alleles of *that* gene (and hence no different phenotypes) available to choose from for breeding purposes; only one version (allele) of that gene (and one phenotype) is available.

  2 The particular *mixture* of alleles in a gamete (sperm or egg) depends on the random shuffling of genes (alleles) that occurs at meiosis. The technical term for the random shuffling of alleles at meiosis is *recombination.* Two processes can contribute to recombination: *crossing over* between homologous chromosomes (chromatids); and *independent assortment* of chromosomes.

  3 There is randomness in fertilization, i.e. which particular sperm fertilizes which particular egg.

To ensure that you appreciate these points about the randomness of inheritance, particularly the shuffling of genes, you should refresh your memory by reading the following box.

Box 2.1 The random shuffling of genes (alleles)

2.1

Put simply, a **gene** is the *inherited instruction* for a particular characteristic. Such a characteristic is called a **phenotypic character** or **phenotype** (though this latter term can also be applied to the organism as a whole). So there are specific genes, say, for flower colour in plants, other genes for petal shape; genes for wing shape in flies; or whatever. So each organism inherits many genes, each gene specific for a particular phenotypic character in that organism. Physically genes are specific stretches of **DNA**, and each particular gene has a particular location along the length of a DNA molecule. In eukaryote cells, DNA is contained within **chromosomes**, structures visible under the microscope at certain stages of the *cell cycle*, the cyclic process of cell growth and division (see Box 4.1, later).

Changes can occur to genes due to what is known as **mutation**. These altered, **mutant**, genes can be passed on, i.e. inherited. So each particular gene, at its specific location in the DNA, can occur in one or other of a number of different versions. Thus each version of a gene arises from mutation, perhaps far back in its ancestral past, many generations of organism ago. The instruction will be slightly different in each version of a particular gene. Each version will therefore instruct an altered phenotypic character; a fly with curly wings rather than straight ones perhaps. Each different version of a gene is called an **allele** of that gene.

Every organism inherits many genes, and so has many phenotypic characteristics. Consider humans. Unless you are an identical twin (i.e. a *monozygous* twin), one from a single fertilized egg, it is extremely unlikely that you look exactly like any of your brothers or sisters. Yet it is likely that you do share *some* physical characteristics; have your father's nose or your mother's blue eyes in common, for example. This, of course, reflects that you and your brothers and sisters have all inherited characteristics (via alleles of genes) from the same parents but that the particular *mix* of characteristics (via alleles) differs. This comes about from a combination of factors. Firstly, except for monozygous twins, you each derive from a different egg and a different sperm and there is randomness in which sperm fertilizes which egg. Secondly, it is very unlikely that any two sperm or any two eggs (**gametes** or **germ cells**) have exactly the same mix of alleles. This extremely low probability of any two gametes being genetically identical arises from the mechanics of **meiosis**, the process by which gametes are produced.

Put briefly, in meiosis the number of chromosomes typical of the body (**somatic**) cells is halved to produce gametes. Thus a human gamete has 23 chromosomes (the **haploid** number) as against 46 (the **diploid** number) in a somatic cell; in the broad bean an egg or pollen cell (the gametes in a plant) has 6 chromosomes, and a somatic cell has 12.

Moreover, the halving is specific. In effect, the diploid cells from which the gametes are derived have pairs of chromosomes (**homologous pairs**) and *one* member of *each* pair ends up in a gamete. But for any one homologous pair, *which* member ends up in any particular gamete is *random*. Therefore there is no special likelihood of a *particular* member of one homologous pair ending up in a gamete with a *particular* member of any other homologous pair. The chromosomes (of homologous pairs) are said to undergo **independent assortment**.

Let us consider independent assortment for just two different homologous pairs of chromosomes, as shown below. (One member of each pair is shaded pink, the other red.)

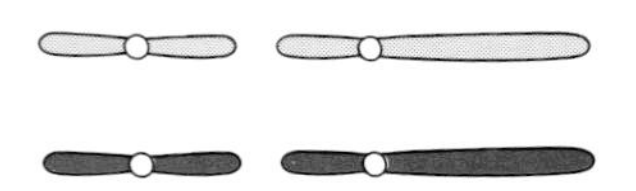

(We can ignore here the fact that *each* member of a homologous pair is in fact comprised, at particular stages in meiosis, of two *identical* 'chromosomes', or **chromatids** as such are called. It has no numerical effect on the following calculation of the consequences of independent assortment.) Now, if each chromosome of a pair assorts independently of any other, the possible combinations in the gametes are:

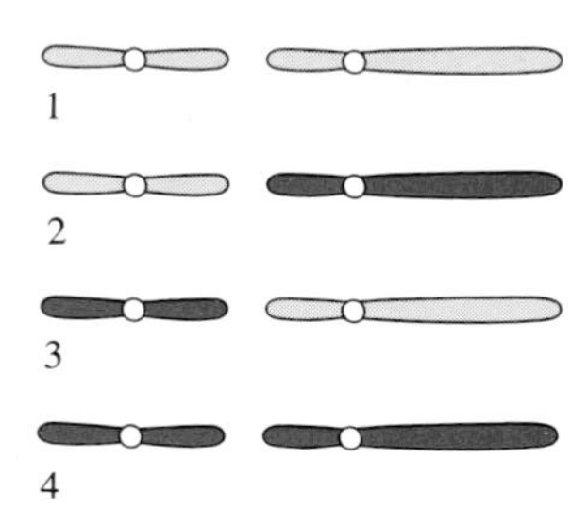

Hence there are four different possible types of gamete from just two different homologous pairs and therefore a 1 in 4 chance of any two gametes having the same combinations of chromosomes.

If we extend this analysis to humans we find that independent assortment of 23 homologous pairs of chromosomes means that there is just *one* chance in 8 388 608 that any two gametes produced by the same individual will have the same combination of *chromosomes*. As genes (alleles) are on chromosomes it is therefore very unlikely that the same mix of alleles will occur in two different gametes.

But the number of possible different gametes brought about by independent assortment is only the tip of the iceberg of the genetic shuffling that occurs at meiosis. Remember that the separation of chromosomes, giving rise to independent assortment, is a relatively late event in meiosis. At the beginning of meiosis, each chromosome in a homologous pair actually comprises two identical chromatids. *Prior* to separation of the chromosomes, adjacent chromatids, one from each of the homologous chromosomes, can cross over and exchange material (see Figure 2.3).

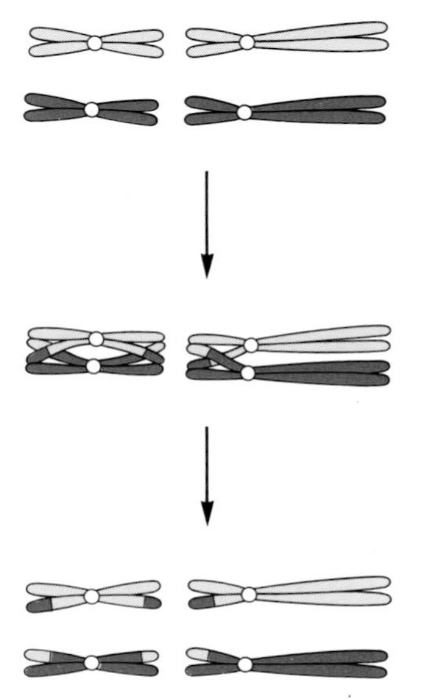

Figure 2.3 Crossing over of chromatids. For simplicity, just two pairs of homologous chromosomes (each comprising two identical chromatids) are shown. Note that there is a *double* cross-over in the smaller pair.

Where along the chromatids **crossing over** occurs is random. Thus two genes linked together on the same chromosome can be separated if crossing over occurs *between* them. If it does not then they will assort together as part of the same chromosome, the same **linkage group**. Crossing over can generate a vast number of *new combinations of alleles* arranged along the lengths of the chromatids. Couple this shuffling of alleles that occurs via crossing over to the shuffling that is introduced *later* by independent assortment of chromosomes and the number of possible genetically different gametes is quite astronomical. Given all this shuffling of alleles at meiosis (known as **recombination**) plus the randomness of fertilization, it would indeed be extremely surprising if any two *dizygous* siblings (i.e. from two different fertilized eggs), human or otherwise, were genetically identical. ■

To test your understanding of the material in Box 2.1 you should now attempt the following question.

Question 2.1 Fill in the blanks in the following passage.

A species of rodent has a ______ number of 14, this being the number of ______ in its gamete cells. Its ______ number is therefore 28, this being the number of ______ in its ______ (i.e. body) cells. Varieties of the rodent are known, varying in ear length and/or eye colour. There is a ______ for each of these two characteristics and these two ______ (for ear length and eye colour) exist on different ______ and so assort ______ during

______ (i.e. the production of gametes). Two different ______ of the ______ for ear length are known; one responsible for long ears, the other for short ones. Likewise there are two different known ______ of the ______ for eye colour; one responsible for red eyes, the other for white eyes.

Where a desirable variety of animal or plant has been obtained by selective breeding, inbreeding can help maintain that variety. However, if we wish to obtain new varieties we may have to resort to *outbreeding* — that is, cross-breed two different varieties in an attempt to obtain offspring with the best characteristics of each of the two parents. But even where desirable parental characteristics may be available, the shuffling of alleles at meiosis and random fertilization limit severely our degree of control over what mix of characteristics any offspring will exhibit — you might get some desired characteristics in the offspring but some less desirable ones along with them. Breeders also need luck to avoid what we might term the *Isadora Duncan effect*. So the story goes, that beautiful and innovative dancer once suggested politely to playwright George Bernard Shaw that they procreate together, as with *her* beauty and *his* brains what truly wonderful children would result. Shaw declined, apparently equally politely, querying what might happen should the mix be the other way around. For would-be breeders the spectre of randomness of meiosis and fertilization raises its ugly (or brainless?) head.

Indeed in the 1920s the cross-breeding of a radish with a cabbage (two closely related species that can be cross-bred) yielded a hybrid plant with radish-like leaves and cabbage-like roots. In fairness to the achievements, or at least the perseverance, of plant breeders, we must add that more recent attempts at cabbage–radish crosses have been more successful. The more recent results, known as Raphano-brassicae, are regarded as promising crop plants.

Look back at the second paragraph of this chapter. The significance of the first word italicized should be evident. *Specific* genes can be transferred from one organism to another. As genetic engineering side-steps normal sexual reproduction, the lottery of meiosis and fertilization can, likewise, be avoided. In principle, just the required gene or genes, and no others, can be transferred from one organism to another. How this is done forms the basis of Chapters 4–8. There is the additional advantage that side-stepping sexual reproduction can save considerable time — there may be no need to wait for an organism to reach sexual maturity before it can act as donor, or recipient, in gene transfer.

The other major natural barrier affecting plant and animal breeders is the **species barrier**. That is, sexual reproduction can only occur between organisms of opposite sex *within the same species*.

▷ What is the definition of a **species**?

▶ A population of organisms that can interbreed to give *fertile* offspring, *under natural conditions*. Conversely, two populations that are *reproductively isolated* are members of two *different* species.

So this means that a barrier to interbreeding, a *species barrier*, is at the very heart of the definition of what a species actually is. However, in some types of organism the definition of a species, and therefore the species barrier, is not quite so clear-cut; notably, some species of plants can, under particular circumstances, cross-breed with different closely related species to give fertile offspring.

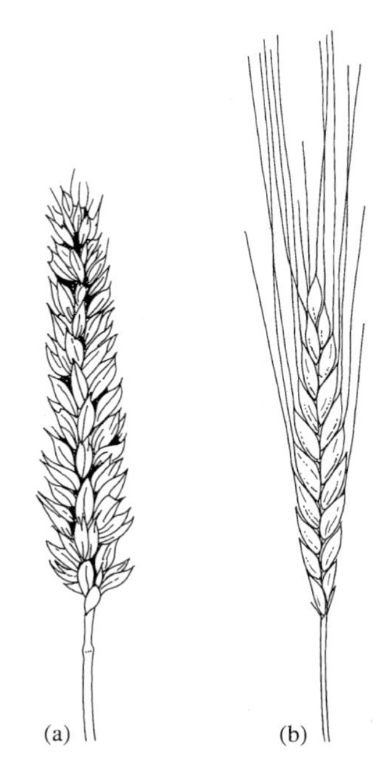

Figure 2.4 (a) An ear of a modern variety of bread wheat. (b) An ear of an emmer wheat.

It is probably to one such accident that we owe one of our major staples — bread. It seems that some 8 000 years ago, in what is now western Iran, a farmer planted his emmer wheat (*Triticum turgidum dicoccum*, very closely related to the durum wheat still grown today for macaroni flour; Figure 2.4b). It happened to be near a patch of one of its relatives, a wild grass (*Triticum tauschii*). A chance fertilization of the emmer wheat by the wild grass led to a new species, *Triticum aestivum*, that we know as bread wheat, or more commonly just wheat. World wheat production in 1989 was over 538 000 000 tonnes of which the majority was *Triticum aestivum* (Figure 2.4a; see also Figure 2.2).

▷ In contrast to this accidental breeching of the species barrier to give bread wheat, can you give an example of a deliberate attempt to breach the barrier to produce a new useful variety?

▶ You will obviously have thought of the successful cabbage–radish hybrids (Raphano-brassicae) or the less successful earlier attempt (that is, less successful unless you like cabbage roots and radish leaves!). There are a number of other possible examples. Triticale, a modern cereal crop, is a hybrid between wheat and rye; the tangelo is a successful fruit, from a cross between a tangerine and a pomelo (a grapefruit-like fruit).

Notwithstanding such exceptions among certain plants, we may take it that in general a species barrier prevents breeders cross-breeding between different species to give fertile offspring. This inevitably limits the variety of offspring possible; a hyacinth-blue rose might be sought but this cannot be attempted by cross-breeding a blue hyacinth and a rose, two species that are sexually totally incompatible.

But as we implied at the start of the chapter, unlike classical breeding techniques, *genetic engineering* not only allows specific gene transfer but does so *across the species barrier*. This is indeed one of its most important features, as you shall see throughout this book. It has already been put into practice by plant and animal breeders, though the results are not yet of commercial importance. For example, *tomato* plants have been made resistant to a specific herbicide by transferring a gene from a *bacterium* into them, potentially allowing you to kill weeds with the herbicide without affecting the tomato plants; *tobacco* plants resistant to predator moths have been constructed by transferring in another gene from a *bacterium*; *sheep* that produce a human blood-clotting factor in their milk have been engineered by transferring in the appropriate *human* gene. And yes, the blue rose is now with us (Figure 2.5).

Petunia genes into roses, bacterial genes into plants, human genes into sheep—the species barrier can indeed be widely breeched via genetic engineering; in principle, genes from *any* organism can be transferred into *any* other organism. Most significantly, genes from a variety of species—animal, plant and bacterial—can be transferred into bacteria and other microbes, creating novel 'microbial factories'.

Activity 2.2 *You should spend up to 15 minutes on this activity.*

At the beginning of this chapter you were told that you would be expected to produce a summary of its *main points* after you had studied it. Now is the time to do so. Your summary should be in the form of a list of numbered points and can be either in the order in which these points appear in the chapter itself, or in a different order if this proves better suited to bringing out those points. The summary should be brief, amounting to no more than 250 words, or much less if done in note form. You need *not* include material in Box 2.1 in your summary.

Thorny problem of blue rose solved

By Geoffrey Lee Martin in Sydney

A HOLY GRAIL of the horticultural industry, the blue rose, is now a reality, according to a team of Melbourne scientists.

Spurred on by market researchers who say blue is the world's favourite colour, they have spent four years isolating the gene responsible for blue shades in petunias, irises and delphiniums.

The breakthrough promises to bring a bonanza for Calgene Pacific, a plant biotechnology company, and its partner Suntory, of Japan, with roses worth £350 million of the world's £15 billion cut-flower market.

Dr Edwina Cornish, the project's principal research scientist, said the blue roses would initially go to the exclusive Japanese gift market, where they could fetch £50 a stem. The first blooms are expected next July, with commercial trials early in 1993.

Calgene has filed for patent protection, effectively putting it ahead of its rival researchers, the US-based DNAP group, and Florigene of Holland.

'Our research shows people do want blue roses', said Dr Cornish. 'The cut-flower market is driven by novelty. It's like the quest for the black tulip.'

She said that because roses lack the pigment needed to produce blue blooms, conventional breeding methods are useless. Instead, the research team isolated a blue gene in petunias that can be transferred to other flowers.

While white roses would seem the obvious recipients for the gene, Dr Cornish said they often lacked a 'pigment production pathway'. She added: 'We may have to do some fine-tuning with other colours so that the pigment gives the optimum blue — that's why they won't go on sale for 18 months or so'.

Chrysanthemums, carnations and gerberas are likely to follow roses down a turquoise path.

Fred Whitsey, Gardening Correspondent, writes: If such a rose passes stringent tests for resistance to disease, form of flower and the quantities in which the blooms appear, which all new roses must undergo, it could end a search that has gone on for centuries.

Efforts to produce a blue rose reached a climax in the 1960s. Despite a substantial part of plant-breeders' research being devoted to this end, and many thousands of seedlings being raised and discarded, the prize eluded them.

They succeeded only in developing roses in various tones of sickly mauve.

They abandoned their attempts, concluding that if a blue rose did one day appear it would come either spontaneously or by laboratory work done outside the sphere of those making crosses, however intelligently inspired, by transferring pollen from one variety to another.

Moorish gardeners, however, had achieved blue roses, including one said to be the colour of lapis lazuli, by treating the roots with indigo dye in the way that carnations are sometimes made to change colour.

Figure 2.5 From the *Daily Telegraph*, 15 August 1991. Genetic engineers achieve what traditional plant breeders cannot. (Incidentally, this success occurred years after I had first thought of the blue rose as a hypothetical example of a possible (?), but perhaps unlikely, target for genetic engineering — a case of genetic engineering converting fantasy to reality; though admittedly using petunia, not hyacinth, genes. Clearly, in future I shall have to try harder in chosing 'ludicrous' examples!)

Question 2.2 Which of the following statements are true and which are false?

(a) Despite the very heavy odds against it, if two eggs in a female had precisely the same mix of alleles then there would be a high probability of their leading to genetically identical individuals following fertilization and development.

(b) Though genetic engineering can breech the species barrier, it cannot cross between taxonomic kingdoms.

(c) Tomato plants have been rendered resistant to a particular herbicide by transferring a bacterial gene into them.

Question 2.3 Give three examples of animals or plants for which traditional selective breeding has been employed.

3 Novel microbial factories

This chapter is longer than Chapter 2. This is mainly because, before considering genetic engineering, it has to introduce a number of concepts about growth and reproduction in microbes that may be less familiar to you than the equivalent processes in farm animals and crops. Just as for Chapter 2, you will be asked to prepare a summary of the main points of Chapter 3 when you have finished studying it (in Activity 3.3).

Our use of animals and plants for food, clothing, shelter, transport, sport or whatever is by no means new—some aspects being as old as humankind itself. As discussed in Chapter 2, this has long included selective breeding of such organisms to better suit our needs. But as well as the familiar animals and plants, other less obvious living organisms, *microbes* (also known as *micro-organisms* or commonly, and loosely, as *bugs*) have also been much exploited.

3.1 A long history of exploitation

As the word implies, a **microbe** is a living organism that cannot be seen with the naked eye. This is a purely functional definition and has no particular taxonomic or evolutionary meaning. Thus as a whole, 'microbes' encompass a wide range of living organisms, some related only very distantly. Some microbes—the bacteria and blue–green bacteria (also known as blue–green algae or cyanobacteria)—are **prokaryotes** and as such have no nucleus or other internal organelles. Other microbes do, and are thus **eukaryotes**. Among eukaryote microbes are numerous fungi including both yeasts and moulds, protista (e.g. *Amoeba*) and algae. (Not all fungi and algae are microbes; some can be seen readily without the aid of a microscope. After all, mushrooms are fungi and seaweeds are algae.) Many microbes such as yeasts and bacteria are single-celled (i.e. **unicellular**) organisms while other microbes are **multicellular**. And some microbes, **viruses**, are not cellular at all! It is only inside the living cells of other organisms that viruses can reproduce. Thus viruses are very small. Unlike other microbes, they are too tiny to be seen with the light microscope and their visualization had to await the invention of the electron microscope in this century. In this book, where we use the word 'microbe' you may assume that we are referring to *unicellular* organisms such as prokaryote bacteria or eukaryote yeasts. Where it is important to distinguish other types of microbe, such as viruses, we shall do so explicitly.

Figure 3.1 Apparently the 'medicinal' value of alcohol has been long recognized. This 4000-year-old Sumerian tablet is a doctor's prescription. A variety of plant extracts were used, as were bits of birds, cows and snakes. *Kash-e U-tu*, 'dissolve in beer', is a phrase found more than once. (Sumer was in what is now Iraq.)

Despite their invisibility prior to Anton von Leeuwenhoek's 17th century microscope, microbes have a record of service to human society almost as long as that of the more familiar animals and plants (Figure 3.1).

Rotten fruit or grain ferments. On occasions the product is much more inviting than that description implies. No doubt this fact was exploited in the early production of wine and beer, but all was dependent on the fortuitous occurrence of favourable microbes, such as yeasts, on the skins of fruit or the husks of cereal. But the products of such happy accidents were highly prized and so there was a considerable motivation to develop more sophisticated and reliable fermentation techniques.

It appears that by 1500 BC, Egyptian beer makers actually *added* relatively pure yeast to their brews. (Though individual unicellular yeasts were invisible, lumps of yeast containing billions of organisms were not.) Care was also taken to control the extent of fermentation. For example, wine was stored in sealed jars, as was beer (Figure 3.2), to prevent its further conversion to vinegar, a process that is dependent on a second fermentation by another, invading, microbe, the bacterium *Mycodermata aceti*. Unless of course vinegar was required, whereupon presumably the wine jars were left unsealed. Vinegar was important both medicinally and as the strongest acid solvent then available.

▷ Can you name another ancient solvent produced by fermentation?

▶ *Kash-e U-tu* (Figure 3.1) implies that alcohol, as present in beer, was put to this use in Sumer.

Figure 3.2 Baking and brewing in Egypt around 2400 BC. The top panel shows the preparation and grinding of the grain (probably emmer wheat or barley). The middle panel shows the malting, kneading and baking of the 'beer-bread'. The bottom panel shows the straining and fermenting of the mash, with the beer being stored in jars which were capped and sealed with clay.

A further consequence of using yeast, the production of bubbles (of carbon dioxide gas), did not go unnoticed or unexploited. By about 4000 BC, the Egyptians were using this fact to bake leavened bread. Baking and brewing were thus often associated trades and among the most technologically sophisticated of the ancient world (Figure 3.2).

The common origin of baking and brewing is still to be seen among some Egyptian peasant workers. When travelling, the fermented dough can be conveniently carried and left with water, whereupon an alcoholic liquor is produced and can be poured off it. This 'beer-bread' beverage is known as *booza*—a long history of drink encapsulated in a single word, perhaps? It's a pity to spoil a good story, but most dictionaries have it that our word 'booze' reaches us other ways; though far from certain, it may derive from old German or Dutch words *buise*, a large drinking vessel, or *bûs*, a blown-up condition!

The Middle East was not the sole seat of the fermentation industry; bread and booze were not the sole products; yeast was not the sole microbial benefactor. The production of the Mexican drink *pulque* by fermentation of the agave plant involves a bacterium, *Zymomonas mobilis*. Milk has long been 'stored' as yoghurt, which also requires fermentation by bacteria, ones that produce lactic (from *lactis*, Latin for milk) acid. A variety of moulds and bacteria are used in cheese making. In Japan, in addition to *saké* (rice wine), there has been a long development of sophisticated technology in producing a variety of fermented foods such as *miso* (bean paste), *shoyu* (soy sauce) and *tofu* (bean curd). These and many other fruits of fermentation are today still an important part of the diet of a variety of cultures worldwide. A glance at Table 3.1 reveals some interesting features and a great deal of variety, even though it represents just a very few of these products:

- There is a wide range of raw materials (grain, pulses, fish, milk, fruit, cabbage, honey and so on), microbes and products involved.
- The microbes involved include both prokaryotes and eukaryotes.

▷ Can you select a prokaryote and a eukaryote from Table 3.1?

▶ There are many of each to be found in the table. All the bacteria mentioned, such as *Zymomonas mobilis* (see pulque) or *Lactobacillus sake* (see saké) or *Lactobacillus brevis* (see säuerkraut and tongbaechu kimchi) are prokaryotes; whereas the various moulds and yeasts (e.g. *Saccharomyces cerevisiae*; see beer), being fungi, are eukaryotes.

- Many products classed under a single general name can exist in a number of significantly different varieties; cheese is an example shown, of which just a few of the many known varieties are listed.

Table 3.1 Present-day fermented food and drink worldwide.

Product	Country	Main raw material	Description	Some major microbes
beer	worldwide	barley	alcoholic drink	*Saccharomyces cerevisiae*
booza	Egypt	wheat	alcoholic drink	probably yeast and lactic acid bacteria
belachan	Malaysia	shrimp	flavouring	various bacteria
camembert	France	milk	cheese	*Streptococcus lactis, Streptococcus cremoris, Penicillium camemberti, Penicillium candidum*
cheddar	UK	milk	cheese	*S. lactis, S. cremoris, Streptococcus durans, Lactobacillus casei*
chicha	South America	maize	alcoholic drink	various moulds, yeasts and bacteria (e.g. *Aspergillus* sp., *Saccharomyces* sp., *Lactobacillus* sp.)
gruyère	Switzerland	milk	cheese	*S. lactis, Streptococcus thermophilus, Lactobacillus helveticus, Propionibacterium shermanii*
idli	India	rice and gram (a pulse)	steamed cake	*Leuconostoc mesenteroides, Streptococcus faecalis*
injera	Ethiopia	tef (a cereal)	pancake-like bread	various moulds, yeasts and bacteria (e.g. *Aspergillus* sp., *Candida guilliermondii*)
jackfruit wine	India	jackfruit	alcoholic drink	yeasts
jalebi	India	wheat and dahi (itself a fermented milk)	confection	*S. lactis, S. faecalis, L. bulgaricus, Lactobacillus fermentum, Lactobacillus buchneri, Saccharomyces* sp.
jeotkal	Korea	fish/shellfish	flavouring, side dish	halophilic and other bacteria; yeasts
kecap	Indonesia	soya bean	flavouring	*Rhizopus* sp., *Aspergillus* sp., halophilic bacteria, yeasts
laban rayeb	Egypt	milk	'curd' cheese	*S. lactis, Streptococcus kefir*, and coliform bacteria
muratina	Kenya	sugar cane juice, muratina (sausage tree) fruit	alcoholic drink	probably yeasts
ogi	Nigeria	maize	'porridge-like'	various moulds, yeasts and bacteria (e.g. *Rhizopus* sp., *S. cerevisiae, Corynebacterium* sp.)
pulque	Mexico	agave juice	alcoholic drink	*Zymomonas mobilis, S. cerevisiae, Lactobacillus* sp., other bacteria
puto	Philippines	rice	steamed rice cake	*L. mesenteroides, S. faecalis, S. cerevisiae*
saké	Japan	rice	alcoholic drink	*Aspergillus oryzae, Saccharomyces sake*; other yeasts; other bacteria (e.g. *Pseudomonas* sp., *Lactobacillus sake*)
säuerkraut	Germany	cabbage	pickle	*Lactobacillus plantarum, L. mesenteroides, Lactobacillus brevis*
shoyu (tsoyu) (Japanese soy sauce)	Japan	soya bean and wheat	flavouring	*Aspergillus soyae* or *A. oryzae, S. rouxii, Pediococcus soyae*
tapé ketan	Indonesia	rice	appetizer or desert	*Aspergillus rouxii*, yeasts
tej	Ethiopia	honey	alcoholic drink	*Saccharomyces* sp.
tempe kedele	Indonesia	soya bean	fried or in soup as meat substitute	*Rhizopus* sp. (mainly *R. oligosporus*)
tongbaechu kimchi	Korea	Korean cabbage	pickle	Various bacteria (e.g. *L. mesenteroides, S. faecalis, L. brevis, Pseudomonas* sp.), yeasts and moulds
trahanas	Greece/Turkey	wheat and milk	soup	*S. thermophilus, L. bulgaricus*
yoghurt	widely	milk	'sour milk'	*S. thermophilus, L. bulgaricus*

Note: where the species within a genus is left unspecified we write 'sp.'.

However, in many ways Table 3.1 conceals as much variety as it reveals, if not more. Firstly, and obviously, the table deals only with food and drink; as you will see, microbes may be involved in other useful types of activity (see Table 3.2, later). Secondly, though the table shows the raw materials and major microbes involved, there is no indication of the range and importance of associated technology; how to prepare the raw materials (soak grain, crush or peel fruit, etc.); what microbes, if any, are *added* or what portion of earlier brews; how much water; what temperature and time for fermentation; how to collect or treat the products; and so on. Finally, and most significantly for our current discussion, the table gives no indication of the degree to which the different processes and the roles of the microbes in particular are understood, that is *in the scientific sense*. In this type of knowledge there is very wide variation between one process/product and another.

In some cases the production of the fermented food or drink is still largely a folk craft, much as was the brewing and baking of the ancient world. Each generation adheres to technology, to recipes, handed down over hundreds or thousands of years. You will note that frequently more than one species of microbe is listed as implicated in the production of one particular product. And in many cases the actual microbes responsible are not known or known only in part (frequently the ones we list are just some major 'suspects'). That knowledge is recent, following systematic study of microbes that are present at the start of the process (in the raw materials or in 'starter cultures' from previous fermentations) or that establish themselves from the surrounding environment during fermentation. Adherence to time-honoured recipes generally ensures consistent conditions in which only certain microbes will establish themselves and 'scientific' knowledge, even to the limited extent of knowing that microbes as such exist, is unnecessary. The knowledge needed is that of the skilled practitioner.

These days, in some cases, such as the brewing of beer, *specific* microbes are added consciously. The precise role of a microbe in a process is sometimes well understood. Thus, for example, *Propionibacterium shermanii* bacteria used in making gruyère cheese (Table 3.1) both contribute to the flavour and, in generating carbon dioxide, produce the characteristic holes. Admittedly deliberate addition of microbes can be quite ancient—the Egyptians presumably realized that yeast had an important role. As with all such early history, precisely what the practitioners understood of *how* things worked is largely speculative. But of one thing we can be fairly sure—the ancient Egyptians would hardly have known about microbes in general nor appreciated that their clumps of yeast, though vital to fermentation, actually comprised billions of individual living organisms. Microbes were not discovered till much more recently. An understanding of their 'living status' and their role in fermentation is of even newer vintage.

Although the discovery of microbes followed hard on the heels of the invention of the microscope in the 17th century, it was not until the 19th century that their *living* nature and their role in fermentation was established firmly, notably through the work of Louis Pasteur. Pasteur's contributions as a scientist were numerous, and his work is well-documented. For our present purposes just two of his findings are particularly significant. Firstly, he showed that fermentation was due to *living cells*, and that *different types of fermentation* (giving different products such as wine, vinegar, and so on) could be caused by *different organisms*. Secondly, he finally squashed the long-held view that life could occur by spontaneous generation from inanimate matter.

These two facts combined meant that, to prevent undesirable products, *unwanted* organisms must be kept out of the brew and this could be achieved by killing those present and preventing the entry of others. The *scientific*, *biological*, justification for sealing the wine jars of the ancient world (keeping out *Mycodermata aceti*) to prevent the formation of vinegar had taken a few thousand years arriving. Fittingly, it was from his studies on the spoiling of beer, undertaken on behalf of the brewers of Lille, that Pasteur reached his important conclusions on different fermentation products. But his 'message' to brewers also had ramifications in other food and drink processing industries and medicine too. Sterilization techniques to kill unwanted microbes are now part and parcel of everyday procedure, a fact celebrated in the name of one such technique, *pasteurization*.

It is common these days to read about **biotechnology**, a term coined recently to cover a wide variety of processes. Precisely what it means is somewhat a matter of taste, but broadly it refers to *the provision of goods or services by means of living cells*. This is taken universally to encompass *genetic engineering*, among other areas, and so is seen frequently alongside that term. It also includes fermentation, and sometimes those technologies practised from antiquity are termed *traditional biotechnology* to distinguish them from more modern developments. Such traditional practices, baking, brewing and the like, are certainly *technology* and do depend on living organisms. But in the sense of *understanding* how fermentation works, the role of microbes—the *bio* bit of traditional biotechnology—is relatively recent and can perhaps be said to begin with Pasteur. And it is such *knowledge* of biology that has revolutionized the use of microbes in the 20th century.

3.2 More microbes, more uses

One major input of biological knowledge follows logically from Pasteur's observation that different microbes can give different products—increase the range of microbes studied and you should increase the range of products available. Indeed this has happened. Most significant has been the discovery of antibiotics, products of microbes (notably of the genus *Streptomyces*) which in nature probably serve as microbial 'weapons' against other ('rival') types of microbe. In our hands too, antibiotics have proved powerful agents, against a range of disease-causing microbes. Nowadays, hundreds of different antibiotics, isolated from a variety of microbes, are in use and represent a world market worth tens of billions of pounds per year. The search for new microbes and new antibiotics continues.

Though undoubtedly the most important, antibiotics are not the only new contribution of microbes to the 20th century, as can be seen from Table 3.2 where just a very small selection of the products and uses are shown. Even from such a small sample you will note a quite wide range of products and uses, from food additives and food processing agents to pharmaceuticals, insecticides and plastics, as well as microbes of all types—bacteria, yeasts and moulds. Most of the microbes, as well as their uses, are themselves discoveries of this century. But in one case at least, baker's yeast (*S. cerevisae*), it's a very old friend being put to novel use (Table 3.2).

Finding a microbe with interesting properties is only the first step in its successful exploitation. Good methods for growing the microbe, perhaps in considerable bulk (Figure 3.3), must be developed. Likewise, methods need to be optimized with respect to the particular product(s) or process for which the microbe is wanted. Any such products need to be harvested, and often purified. All of this has benefited from advances made this century in **biochemistry** (the study of the chemistry of living cells), allowing a degree of rational design in developing appropriate techniques. Though, as each microbe and process has unique features, a good deal of empirical, 'trial and error', development also goes on; much as must have been the case over the long history of what are now regarded as traditional products (Table 3.1). And, of course, both the trial and error and the rational design aspects of handling new microbes and new products have led to a more general body of knowledge of technology and biology. Such knowledge can be applied to yet other new products and fed back to the production of more traditional ones too (beer, cheese, etc.).

Table 3.2 A small selection of the 20th century exploitation of microbes.

Organism	Product/use
Corynebacterium glutamicum (a bacterium)	lysine (amino acid used as an animal feed supplement); glutamic acid (amino acid used for monosodium glutamate, MSG, a flavour enhancer)
Aspergillus niger (a mould)	citric acid (food additive)
Xanthomonas campestris (a bacterium)	xanthan gum (polysaccharide used as a thickening or gelling agent in ice cream, mousse, clear gel toothpaste)
Eremothecium ashbyi (a yeast)	riboflavin (i.e. vitamin B_2; pharmaceutical/nutritional use)
Saccharomyces cerevisae (baker's yeast)	invertase (enzyme used in producing confectionary)
Penicillium chrysogenum (a mould)	penicillin (antibiotic)
Cephalosporin acremonium (a mould)	cephalosporin antibiotics
Genus *Streptomyces*, various species (bacteria)	various antibiotics, e.g. streptomycin, tetracyclines
Bacillus thuringiensis (a bacterium)	bacterium itself used as insecticide (see Section 10.2.3 later)
Bacillus subtilis (a bacterium)	α-amylase (enzyme used in making glucose syrups from starch)
Bacillus licheniformis (a bacterium)	protease (enzyme used in 'biological' washing powders and as meat tenderizer)
Alcaligenes eutrophus (a bacterium)	polyhydroxybutyrate (polymer used as plastic; registered as Biopol by ICI)

Note: the scale of uptake of the products/processes shown varies widely.

Figure 3.3 Some amino acids are produced in great bulk (e.g. glutamic acid, Table 3.2, at around 300 000 tonnes per year). As the 20 identical fermenters at KHK's plant in Hofu, Japan show, the size of the 'brews' can rival that of beer—each fermenter is around 30 metres high and holds about 290 000 litres of fermentation broth. They are used to produce glutamic acid and lysine.

As already indicated, one of the great advantages of using microbes as industrial agents is the fact that often they can be grown in large amounts. Obviously, this is necessary where the desired microbial product is itself consumed in bulk. Thus the 1979 world production of 80 000 000 tonnes of beer (around 136 000 000 000 pints!) was accompanied by 1 750 000 tonnes of baker's yeast. Hence microbe and output, beer, can both be considered bulk commercial commodities (but remember that yeast is also needed for bread-making).

What makes all this possible is that, given appropriate conditions, microbes can grow and reproduce extremely quickly. Though some microbes can undergo a form of 'sexual' reproduction, the more usual and rapid mode of reproduction is without sex, **asexual reproduction**. This simply involves a single (*parental*) cell dividing to give two cells (usually termed **daughter cells**). These two daughter cells can then divide to give four, and so on. Thus the total number of organisms (single cells) doubles each *life cycle*, with each cell division. Such growth is said to be **exponential**. The speed at which cells double in number, the *doubling rate* can be very rapid.

▷ In what senses is our description of a single cell dividing to give two a gross oversimplification of the life cycle of a microbe?

▶ The *entire* life cycle of a microbe must involve more than simply reproduction by cell division. Firstly, the microbe must grow or else each round of cell division would reduce the cell size presumably to an ultimate vanishing point! Secondly, as well as growing to double its starting size the parental cell must precisely double (**replicate**) its genetic material, its DNA. Finally, during cell division one complete copy of the replicated DNA, must be parcelled out to each of the two daughter cells.

A schematic view of the asexual life cycle of a unicellular microbe is shown in Figure 3.4.

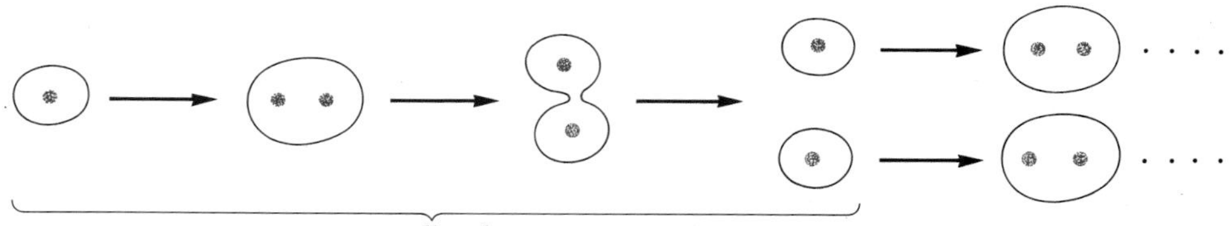

one cell cycle

Figure 3.4 The asexual life cycle of a unicellular microbe. A single microbe (i.e. in unicellular microbes, a single cell) grows to double its starting size and replicates its DNA (shown as a red blob). The large microbial cell then divides to give two daughter cells (i.e. two unicellular microbes), each with a complete copy of the parental DNA. *Each* daughter cell then grows to double its size, replicates its DNA, divides and so on.

Another key feature of microbial growth is that most microbes are superb chemists. Each species of microbe has its own particular optimal diet but many are able to do quite well, that is synthesize all the cellular components needed for growth and division, on a very restricted mix of nutrients. Take, for example, *Escherichia coli*, a generally benign bacterial inhabitant of mammalian intestines, our own included. Outside intestines, and inside laboratory glassware at 35 °C, given a minimal diet of water, oxygen, sources of nitrogen (as ammonium, NH_4^+), phosphorus (as phosphate, PO_4^{3-}), a few other ions and a source of carbon such as glucose, *E. coli* can grow and divide—that is, complete one *cell cycle*—about once every 40 minutes or so. That is, in this time it can convert this frugal diet into all the myriad of different organic substances, some very complex, that constitute *E.coli* itself. And it can do so in sufficient quantity to double its own mass, and allow time for the delicate mechanics of cell division—all in 40 minutes.

Activity 3.1 *You should spend up to 20 minutes on this activity.*

(a) Given an adequate supply of its minimal diet, *E. coli* can grow and divide every 40 minutes at 35 °C. Assuming we start with a single *E. coli*, how many will exist after 6 hours and how many after 8 hours? (You can ignore any cell death that might occur over this period, as this would be very low compared to the production of new cells by cell division.)

(b) Given a richer growth medium (supplemented with some more complex organic compounds), *E. coli* at 35 °C can grow and divide every 30 minutes. Starting with one *E. coli*, how much time elapses before the number of organisms reaches that found in *E. coli* grown as in (a) for 6 hours?

(c) The gestation period of a mouse is about 20 days. *In principle*, how many cells would result from growing *E. coli* for 20 days under conditions like those in (a), starting with a single cell?

As Activity 3.1 shows, the amounts of *E. coli* that can be grown in a short time can be impressive. Though, in some instances, this may be in theory only. Take, for example, the answer to part (c), the around 10^{217} cells that 'would be' produced in about 20 days. As one *E. coli* cell weighs about 10^{-12} g, the total *mass* of cells after 20 days would be around 10^{202} kg (i.e. $10^{217} \times 10^{-12}$ g/10^3). This is considerably more than the mass of a litter of, say, 12 mice produced in the same 20 days. (Note also that, unlike 40 minutes for *E. coli*, 20 days is only the gestation period of the mouse not the full life cycle including the time needed to bring it to sexual maturity.) Anyhow, as 10^{202} kg is also many, many times greater than a reasonable estimate of the total mass of the observable Universe (based on an estimated 10^{87} electrons), the number of *E. coli* 'achieved' is most definitely *in principle* only—presumably the cells would run out of glucose well before then. However, this calculation amply demonstrates the rapidity of growth of microbes as compared with animals (e.g. mice) and plants.

E. coli is not even the fastest-growing microbe. For example, under ideal nutritional conditions the bacterium *Bacillus stearothermophilus* has been timed to double in 8.4 minutes, *E. coli* staggering in at a pathetic 21 minutes.

This rapid rate of reproduction and the wide variety of microbes and hence microbial products available has meant that more and more we have come to regard microbes as 'factories'. Choose the microbe that gives the desired product, optimize the conditions for its growth and output of the product, harvest the product.

In a sense the exploitation of a wider range of microbes is like the farmer's domestication of animals or the cultivation of what were wild plants. Animal husbandry and crop management have their equivalents in fermentation technology—finding the best economic diet on which to grow a particular microbe, harvest its product(s) and the like. The third strand of agriculture—the contribution of animal and plant breeders from antiquity to the present day (Chapter 2)—also finds a parallel among microbes; and our use of microbes has benefited from a knowledge of genetics.

3.3 Breeding better 'bugs'

However remarkable the chemical virtuosity and rapid reproduction of a microbe, it is unlikely that as first isolated from the wild, it cannot be improved upon. Of course 'improvement' is highly subjective and our standpoint is that of the user, not that of the microbe! Can the microbe be grown more rapidly or on cheaper nutrients, can the output of its desired products be increased...? These are the kinds of questions likely to be posed by potential users. We have already referred to one type of solution—trial and error juggling (aided by experience and scientific knowledge) of the conditions in which the microbe is grown until the optimal ones are discovered. Another possible solution is to 'juggle' the microbe. Let us take an example.

Commercially, alcohol in most beer is produced by the metabolic activity of baker's yeast, *Saccharomyces cerevisae*, acting on glucose as a source of carbon.

▷ The first part of the production of alcohol, the **metabolic pathway** leading from glucose to pyruvic acid, is common to virtually all living organisms. Can you name this pathway? What happens to the pyruvic acid?

▶ **Glycolysis** is the name of the pathway. What happens to the pyruvic acid produced by glycolysis depends on the organism in question and the conditions. In organisms under aerobic conditions, pyruvic acid, is further metabolized to carbon dioxide and water. This latter stage requires oxygen to operate. So in the absence of oxygen, i.e. under *an*aerobic conditions, pyruvic acid has other fates. In anaerobic yeast this fate is to be converted to alcohol.

However, alcohol is toxic to yeast and as the alcohol in the brew increases in concentration it eventually kills the yeast that produces it. Finding or 'developing' a yeast that is relatively alcohol-tolerant allows the brewing of stronger beers. In principle, what is done is much as did the ancient maize growers of Mexico who probably kept back some of the best cobs from which to grow next year's crop. Just as there is genetic variety among maize, so there is among the billions of individual organisms in a batch of yeast. The trick is to select the best individual yeast (in this case, the most alcohol-tolerant) from which to breed a whole new batch of yeast for future use. There is the advantage with microbes that having selected the individual it takes next to no time to get large numbers of its descendants; another bonus from rapid reproduction (Activity 3.1).

▷ Say you were a plant breeder trying to produce new useful varieties of a particular species of plant. What two avenues for getting genetic variation would be available to you?

▶ The *ultimate* source of all genetic variety is changes in DNA, mutations, which give rise to different versions (i.e. alleles) of genes (Box 2.1). So one avenue is to search among members of the species in question for hitherto undetected and hopefully useful alleles. Having found them you can then attempt to breed them into your stocks. But for sexually reproducing plants a further route to variety is open to you. You can cross-breed different varieties of the species in the hope that the huge number of different possible *combinations of alleles* that are produced via recombination at meiosis, and random fertilization (Section 2.1; Box 2.1), will include new useful combinations. (The same two strategies would be also applicable to animals.)

▷ Which of the two above sources of genetic variation are available to breeders of microbes that reproduce only asexually?

▶ As there is only a *single* parent (a single cell) and reproduction is asexual, just the same set of genes is copied and passed on generation after generation. Therefore there can be no variation introduced by reproduction as such. However, during the replication of the DNA, cell cycle after cell cycle, mistakes will arise; such mistakes, *mutations*, can give rise to genetic variants (mutants; new alleles) with new characteristics.

Of course where in the DNA, in which genes, mutations arise is random, a matter of chance (Section 2.1). So just like the animal and plant breeder, the microbe breeder still depends on chance—that a desirable mutant, say a more alcohol-tolerant individual yeast, will crop up and that this can be selected to produce similar descendants in large numbers.

But here the microbe breeder has an advantage, again dependent on rapid microbial reproduction. The desired mutation may be a very rare event but, given the huge numbers of individual microbes that can be produced in a short space of time, the breeder has a huge choice and hence a reasonable chance of finding just such a mutant.

For example, suppose there is a chance that just one individual in a million will have a desired mutation. Now consider the problem for a microbe breeder who can produce billions of individual organisms overnight (Activity 3.1). There are likely to be thousands of the desired mutants (i.e. one millionth of billions) among them. So the problem is not so much how to get the desired mutants but more how to find them—how to select the desired individual(s) from the 'microbial haystack'. Unlike a desirable cob of maize, how do you spot a desirable invisible individual microbe? There are tricks to do this, which we will mention in Chapter 6; suffice it to say for now that it can be done, often quite readily.

In contrast, for the animal breeder, spotting the mutant—say a cat with 'desirable' coat colour—may not be the problem. But if the mutation here also has a one in a million chance of occurring, then actually getting such a mutant will be the problem—it's not so easy to obtain millions or billions of cats. And if the desired mutation hasn't occurred, then you can't select it to breed from—a small '(cat) haystack' without any needle! In their search for desirable mutants, plant breeders can obtain and scan much larger numbers of individuals than can animal breeders. But here too, the actual numbers scarcely rival the 'almost limitless' numbers achievable with microbes (Activity 3.1).

The intrepid mutant-microbe hunter can also increase the chances of success by increasing the *frequency* of mutation, i.e. treat a batch of microbes with agents that cause mutations.

▷ Can you recall any such agents?

▶ Certain chemicals and forms of radiation (ionizing and non-ionizing) can cause changes in the DNA, i.e. can cause mutations. Such agents are called **mutagens**.

Using mutagens is a far from specific process. Where in the DNA molecule mutations occur is still random. But producing more mutants (i.e. more per given population than in the untreated case) widens the choice and so increases the chances of desirable ones being found. Of course, animal and plant breeders can also use mutagens to increase the frequency of mutation. But, for reasons that we need not consider, using mutagens on microbes is easier.

Thus, for a number of reasons, obtaining useful new mutants is easier in microbes than in animals and plants. There is, however, one source of producing more variety open to animal and plant breeders that is unavailable to breeders of asexually reproducing microbes.

▷ What is this?

▶ Sexual reproduction (see above and in Section 2.1).

Sexual reproduction allows a breeder to take available mutants (alleles) and produce new varieties of organism by creating *new combinations* of mutants (alleles). Thus, for example, a mutant thornless rose may turn up, as may another one that has a very strong perfume. Cross-breeding the two varieties *might* yield offspring with the desired mix of no thorns and strong perfume.

▷ Why the '*might*'?

▶ Remember the genetic lottery of sexual reproduction (Section 2.1; Box 2.1).

Nevertheless, sexual reproduction does offer a *chance* of producing varieties with useful new combinations of alleles—this opening is not on offer where reproduction is totally asexual. But some microbes, including baker's yeast, can reproduce both asexually *and* by what amounts to a form of sexual reproduction. So in such cases the microbe breeder could also exploit this sexual cycle of reproduction to get new varieties of microbe; new combinations of alleles. And having perhaps got a useful new variety via the sexual cycle, the breeder can then exploit the rapid *asexual* cycle to produce a large population of the variety in next to no time—the best of both worlds.

Activity 3.2 *You should spend up to 15 minutes on this activity.*

Consider what you have learnt about microbial growth and reproduction. Assume that we are dealing with a particular microbe that can undergo only an asexual cycle of reproduction. Is it true to claim that 'producing new genetic varieties of the microbe can be more specifically controlled by the experimenter than can the equivalent process in an animal or plant'?

Give reasons for your answer and write them down as numbered points in note form, in no more than about 200 words.

3.4 Enter genetic engineering

Just as traditional techniques have yielded improved varieties of animals and plants, so has their 'equivalent' (selecting mutants) succeeded for microbes. For example, the first commercial production of penicillin was achieved by taking some of the mould *Penicillium chrysogenum* (Table 3.2) and putting it through 21 successive stages of mutation. This yielded a new **strain** (i.e. a variety within the same species) that yielded 55 times more penicillin than the original one. (Nowadays, further improvements of strains and in fermentation technique, mean that yields are even greater than in the early war-time production.) Nevertheless, such procedures are laborious and uncertain—perhaps many rounds of selecting mutants and each round a gamble: will it yield a better strain or not? The *relative* certainty of genetic engineering, i.e. the ability to transfer *specifically*, just a desired gene and no other, is obviously attractive. (Of course to do this, such a gene still has to be available; but wherever it occurs, irrespective of species, it can be transferred.) Let us take another example, once again from the traditional biotechnology of brewing.

As outlined earlier, the alcohol in beer derives from the metabolic activity of the yeast *S. cerevisae* on glucose. But glucose is too expensive a material to be fed to yeast for this purpose. What is used is a glucose-containing polysaccharide, starch, in the 'impure' form of suitably treated grain (usually barley). But *S. cerevisae* does not possess enzymes capable of degrading starch to glucose. So before adding yeast, the grain must be treated with starch-degrading enzymes, as present in malting (i.e. germinating) barley itself. The liquid fraction from this treatment is called *wort*. Adding *S. cerevisae* to this produces the required alcohol. But wort contains a mixture of different sugars, not all of which can be fermented by *S. cerevisae*. Some sugars, *dextrins* in particular, remain unfermented. Commercially this leaves room for improvement; because in certain countries, including a 'calorie-conscious' USA, there is a desire for low carbohydrate, 'lite', beers (i.e. with less residual carbohydrate). So yeast capable of breaking down dextrins would be of obvious advantage. There is little chance of mutating *S. cerevisae* to achieve this end. The reason is that while mutation may by chance eliminate or alter existing enzymes, it is very much less likely to produce ones totally new to a species (such as desired dextrin-degrading ones).

Saccharomyces diastaticus is a yeast related to *S. cerevisae* and has enzymes capable of breaking down dextrins. And *S. diastaticus* can produce alcohol. So one solution would be to simply switch to *S. diastaticus* for beer production. Unfortunately, such beer is found to have an unpleasant phenolic taste (i.e. like carbolic), more reminiscent of the hospital ward than of the pub.

▷ Can you suggest a role for genetic engineering?

▶ Specifically transfer the gene(s) coding for dextrin-degrading capacity from *S. diastaticus* into *S. cerevisae*.

Indeed we might suggest going even further. Some species of yeast can grow on starch itself.

▷ What does this suggest?

▶ Presumably they possess *all* the enzymes needed for starch degradation.

So, in principle, we might transfer *all* of the relevant genes for starch degrading-enzymes into *S. cerevisae*. How one *might* do so will become evident from Chapters 4–7 (though, in fact, this particular transfer might be quite tricky, as we shall reason in Section 7.4). Obviously the ability to do so will depend on techniques being available to carry out genetic engineering in microbes in general and in yeast in particular. As will be seen in Chapters 4–6, genetic engineering techniques were indeed first developed in microbes, in bacteria in particular, and are easier to perform here than in animals and plants.

Clearly, the application of genetic engineering to microbes to obtain new varieties to enhance existing processes is potentially very useful, whether long-standing processes such as brewing or more recent ones such as, say, the production of antibiotics. Yields could be increased and costs reduced. But it is the application of genetic engineering to get microbes to produce, what are for them, quite novel products that has really generated the excitement.

As we have amply demonstrated, it is the ability to grow microbes and obtain their products in bulk that makes them particularly attractive industrial agents. If this microbial feature can be coupled to the production of other substances which are normally hard to obtain it would be of great advantage—an advantage already appreciated and in a few cases actually realized.

You are probably familiar with one such case—the hormone insulin. This is used to treat diabetics and traditionally pig insulin has been used for this purpose. However this is probably not ideal as chemically it differs slightly from its human counterpart. Thanks to genetic engineering it has been possible to do what *in essence* amounts to the *specific* transfer of a gene (in practice, two genes; see Section 9.3.2, later) coding for human insulin into bacteria and get these bacteria to produce *human* insulin. As the genetically engineered bacteria can be grown in large amounts this should provide a ready supply of human insulin. Approval for sale of such insulin in the UK was given in September 1982.

▷ The transfer of the gene for human insulin into bacteria is *specific* to just that gene. Specificity was a key feature of genetic engineering, as you recall from Chapter 2. What other key feature of genetic engineering does this transfer of the insulin gene demonstrate?

▶ As for animals and plants (Chapter 2), the *species barrier* presents no barrier where genetic engineering *into* microbes is concerned. In principle, genes from any organism can be transferred into a microbe.

Other 'human' products from genetically engineered microbes have come to market and many others are under development. Notable among them are a number of other pharmaceuticals such as interferons (anti-viral and anti-cancer agents). What such products share is that without genetic engineering they would be virtually unobtainable in therapeutic amounts, perhaps not even enough to test their supposed biological effects to see whether they would be really worthwhile producing industrially. With genetic engineering, microbes can be made to 'churn' them out. We shall examine these and other medical products of genetically engineered microbes in Chapter 9.

The use of genetically engineered microbes is not restricted to human products as, in principle, genes from virtually any organism can be transferred into microbes. Nor is it restricted to medicine. For example, the gene coding for thaumatin, a sweet protein isolated from the West African *katemfe* fruit (*Thaumatococcus daniellii*), has been transferred into a number of species of microbe, both bacteria and yeasts. The aim, as yet unrealized, is to get the microbes to produce large amounts of this protein for use as a 'non-fattening' sweetener. Such a microbial source of a plant product is considered to offer an easier, potentially cheaper, and more reliable supply than that from the plant itself.

In essence then, genetic engineering allows us to extend the enormous (natural) range of products and uses of microbes as a whole (e.g. Tables 3.1 and 3.2). We can thus exploit their rapid reproduction and chemical virtuosity still further and create a whole host of new microbial 'factories'—factories performing hitherto novel tasks, producing what are for them 'foreign' products originating in other microbes or plants or animals.

Activity 3.3 *You should spend up to 30 minutes on this activity.*

At the beginning of this chapter you were told that you would be expected to produce a summary of its *main points* after you had studied it. Now is the time to do so, much as was suggested for Chapter 2 (Activity 2.2). The summary should be a list of numbered points and amount to no more than 400 words, or less if in note form. This time your summary should contain brief references to examples illustrating the main points.

Activity 3.4 *You should spend up to 30 minutes on this activity.*

(a) Using the material in Chapters 2 and 3 and/or your summaries of these chapters, make out a case supporting the following statement: 'Genetic engineering promises a revolution in our use of animals, plants and microbes'.

You should write your answer in summary form, that is as a logical sequence of numbered points. You should use proper sentences not just notes. Only include the main supporting points in your answer but where appropriate indicate examples (either from the chapters or elsewhere) that illustrate these points, as brief items in parentheses. You should not spend more than 15 minutes on part (a) and your answer should be in less than 250 words.

(b) Adopting the same style of summary writing as in (a), make out a case supporting the following statement: 'Genetic engineering of animals, plants and microbes is *not* revolutionary but is merely an extension of time-honoured practices'.

You should not spend more than 15 minutes on part (b) and your answer should be in less than 200 words.

Question 3.1 Which of the following are prokaryote and which are eukaryote: baker's yeast; *E. coli*; palm tree; ant?

Question 3.2 Which of the following are obtained by exploitation of naturally occurring microbes: Scotch whisky; oil (petroleum); bread; penicillin?

Question 3.3 Two vats of yeast are set up. One, A, has a relatively poor growth medium wherein the yeast cells divide every 4 hours. In the other vat, B, the growth medium is richer and the yeast divide every 2 hours. If both vats are 'seeded' with the same number of yeast cells (10 000) at the start, which vat will yield the most cells if A is harvested after 12 hours and B after 6 hours?

Question 3.4 Which of the following statements are true and which are false?

(a) Bacteria are attractive to industrial users as large amounts of the organisms can be grown in a short time.

(b) Bacteria are attractive to industrial users as they require simple growth media in which virtually any bacteria can grow.

(c) Human insulin can be produced using genetically engineered bacteria.

(d) One of the main uses of genetically engineered microbes is for medical purposes.

4 Transferring genes into bacteria

This chapter and the four that follow take a different tack from those so far, dealing with specific technical procedures involved in genetic engineering in microbes (and some other organisms) rather than general discussions about why genetic engineering is potentially very useful. On the whole we deal with the main principles of such procedures (plus a few clever 'tricks of the trade') and it is these that you should understand rather than all the details. There is no formal summarizing activity associated with this chapter or the following ones. But by now you may, or may not, feel it nevertheless worthwhile to produce your own summary (with or without illustrative examples and references) as a way of assisting initial absorption of the material and as a useful personal aid to memory when later reviewing it. We give our summary of the chapter at its end.

As we have emphasized, *in principle* it is possible to genetically engineer any organism; to transfer into it the genes from any other organism. *In practice* such gene transfer has already been achieved between a wide variety of organisms—microbes, animals and plants. But *bacteria* were the first organisms into which 'foreign' genes were transferred specifically and it is in bacteria that the techniques have reached their most sophisticated level so far. So we shall explain the main principles of genetic engineering by reference to bacteria and then in later chapters go on to consider some of the problems and advantages of genetic engineering in other organisms. In fact much of our discussion will be derived from work on one particular species of bacterium, *E. coli*, though most of the fundamentals of what we describe are true also for other bacteria.

4.1 The genes of E. coli

As a bacterium *E. coli* is a prokaryote, unicellular and a microbe (Section 3.1). It reproduces asexually by cell division and, given appropriate conditions, can do so very rapidly (Section 3.2). Given its relative simplicity, as compared to a multicellular animal or plant, and our ability to grow bacteria in large numbers, *E. coli* has proved a very useful 'laboratory' organism over the last 40 or so years. Many basic biological features are common to all living cells and *E. coli* has acted as a 'model' for uncovering such universals—the assumption is that what is fundamentally true for *E. coli* is true for *E. lephant*! Even where the links are not so straightforward (see Box 5.2, later, for example), work on *E. coli* has often laid the foundations for studies of other organisms. From such extensive studies more is probably known about the total biology of *E. coli* than that of any other organism. It is thus natural that much of the development of genetic engineering should have been done using *E. coli*.

4.1

Box 4.1 Genes, DNA and chromosomes

There are many definitions of what precisely constitutes a gene and no one definition is totally satisfactory. Perhaps the simplest is that a gene is a bit of heritable information (Box 2.1). The problem then becomes to answer 'bit of information for what'?

At a relatively gross level, some genes can be seen to contain (encode) information for obvious characteristics: for example, a specific gene for pea shape (one version of the gene, one allele, 'saying' round seeds; another allele 'saying' wrinkled seeds).

At a more subtle level of analysis, some genes can be linked to specific subcellular traits, particular genes encoding the information for making particular proteins. Put more precisely, a gene may be that bit of information that codes for the structure of a particular polypeptide chain (see Box 5.1, later). Such genes that code for the structure of polypeptides are sometimes called **structural genes**.

However, other genes do not seem to code for making polypeptides and so are not called structural genes. Among such genes are ones that code for certain specific types of RNA such as transfer RNA (tRNA) and ribosomal RNA (rRNA). **rRNA** is part of the substance of **ribosomes**, the cellular sites of polypeptide synthesis. tRNAs are also involved in polypeptide synthesis, acting to carry amino acids to the ribosomes where they can then become incorporated into the growing polypeptide chain (we deal with this in Box 5.1).

There is a further class of genes that can be defined by their function; these are the genes that control (regulate the function of) other genes.

Another way of answering the question of what comprises a gene is to consider not its 'function' as a bit of heritable information, but of what it is actually itself composed. This has a more straightforward answer—genes are composed of DNA. More precisely, a gene is a *specific* sequence of **deoxyribonucleotides** in one strand (the so-called **coding strand**) of what is a double-stranded DNA molecule, a **double helix**. There are four types of deoxyribonucleotide, distinguished by which particular **base** each contains. It is the *sequence* of bases along a strand of DNA (i.e. each base as part of a deoxyribonucleotide) that encodes the information, much as the sequence of letters in these words encode *their* information. (*Note*: as it is the bases that distinguish one deoxyribonucleotide from the next, one often speaks of a **base sequence**, rather than a deoxyribonucleotide sequence, within DNA.) As a DNA molecule is generally very long, containing many thousands or millions of bases, it comprises many genes. So though a single strand of a molecule of DNA is *physically* an uninterrupted sequence of bases (i.e. each is part of a deoxyribonucleotide and covalently linked to the next deoxyribonucleotide in the *poly*deoxyribonucleotide chain) the continuous sequence is 'punctuated' to give separate genes. This 'punctuation' is itself encoded in the sequence of bases—specific base sequences indicating, say, 'a new gene starts here' or 'this gene ends here' or 'obey this gene only if such and such circumstance pertains'. Yet other sequences of bases in DNA may well indicate nothing! They may not have an informational function at all, being part of what is sometimes called 'junk' DNA.

As will be elaborated in Box 4.2, bases in one strand of a DNA double helix are *paired specifically* with bases in the other strand. Therefore, in speaking of the *size* of a particular DNA molecule (or of a particular gene), this is sometimes seen quoted in terms of the number of **base pairs** that it contains.

In eukaryotes each DNA molecule is bound up with proteins, notably a class of proteins called **histones**. A single long linear molecule of DNA combined with such proteins is compacted, enabling it to fit inside the nucleus of the cell. At its maximum degree of compaction, a state that occurs during cell division (meiosis and mitosis), such a complex of a DNA molecule and protein is thick enough to be visible, as a structure called a chromosome (Box 2.1), when magnified a few hundred times by a microscope. Thus each chromosome in a cell comprises a single very long molecule of DNA, that contains many genes, wrapped up with protein.

In prokaryotes DNA is 'naked' by comparison, as it does not associate with proteins to form 'true' chromosomes. A prokaryote chromosome is nevertheless a long molecule of DNA folded up inside the cell.

(*Note*: the structure of DNA is considered further in Box 4.2.) ■

To test your understanding of the material in Box 4.1 you should now attempt the following question.

Question 4.1 Which of the following statements are true and which are false?

(a) All genes encode the information for determining the structure of polypeptide chains.

(b) In eukaryotes the DNA in the nucleus is bound up with proteins called histones in structures called chromosomes.

(c) In chemical terms chromosomes are large structures and therefore each chromosome must contain many molecules of DNA.

(d) Each chromosome in a eukaryote cell contains many genes.

(e) The information in a gene resides in the sequence of (deoxyribonucleotide) bases comprising that gene.

(f) As a single DNA molecule contains many genes, there must be breaks in the DNA molecule delineating one gene from the next.

In common with other prokaryotes, *E. coli* does not have a chromosome in the sense of 'DNA bound up with proteins', as is the case in eukaryotes. The single chromosome of *E. coli* is composed of DNA without protein. The DNA is itself a single molecule, but in contrast to the familiar *linear* double helix of DNA, the DNA double helix is in the form of a closed circle. This single circular chromosome consists of about 4 million base pairs; i.e. *each* strand of the double helix contains about 4 million bases. (A human chromosome contains a single linear DNA molecule of about 100 million (10^8) base pairs.) When *E. coli* reproduces, it is this single circular chromosome that is replicated, one copy being passed on to each of the two *E. coli* that result from the subsequent cell division (Figure 3.4).

But in addition to its single chromosome, *E. coli* frequently has a number of smaller circles of DNA, also capable of replication. These circular DNA molecules, which in *E. coli* comprise just a few thousand base pairs each, are called **plasmids**. Figure 4.1a is a diagrammatic representation of *E. coli* with its single circular chromosome and a number of smaller circular plasmids; Figure 4.1b shows a high-power electron micrograph of a bacterial plasmid.

Various types of plasmid exist, but, in general, a given bacterial cell, and hence its direct descendent cells, can possess only one type of plasmid at any one time. A cell may however contain many identical copies of that plasmid. Some cells contain tens or even hundreds of identical plasmids.

Not all *E. coli* contain plasmids and so they cannot be vital to the existence of bacteria. All the essential genes are contained in the chromosome and these are passed on at cell division. Plasmids can also be passed on to daughter cells at cell division. There are genes on plasmids too, often ones that encode factors endowing the bacteria with antibiotic resistance. Presumably bacteria with such plasmids have an evolutionary advantage in the presence of other 'hostile' (i.e. antibiotic-producing) microbes. What is important, from the point of view of genetic engineering, is that plasmids can be extracted from bacterial cells. And if one takes some isolated plasmids in a solution containing calcium ions (Ca^{2+}) and adds them to other *E. coli* that have no plasmids,

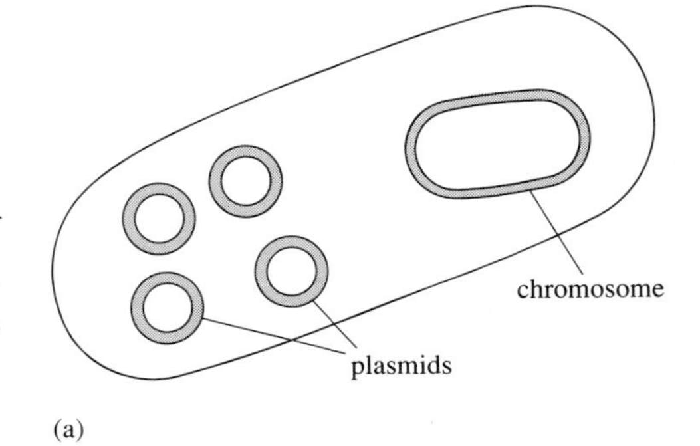

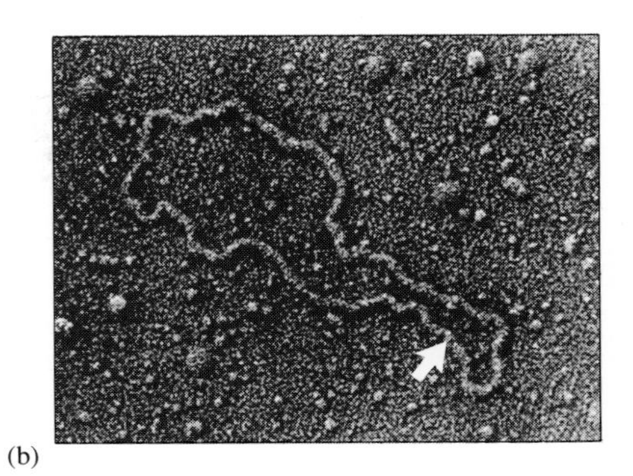

Figure 4.1 Bacterial chromosomes and plasmids. (a) A diagrammatic representation of an *E. coli* showing a single circular chromosome, comprising a double-stranded DNA molecule of about 4 million base pairs, and a number of much smaller identical plasmids (not to scale). (b) A high-power electron micrograph of a bacterial plasmid (arrowed).

these bacteria can take up the plasmids into the cells. (Once inside a cell a single plasmid can soon be present as several copies (Figure 4.1a) via successive rounds of DNA replication.)

Thus plasmid genes can be transferred from one bacterium (with plasmids) to another (without plasmids). That in itself would be interesting but of limited importance.

4.2 The essential steps in gene transfer into bacteria

As well as the discovery of how to transfer plasmids from one bacterium to another, further factors combined to open the way for the development of genetic engineering in the early 1970s. We shall be considering each of these factors in turn, but it is convenient to first summarize them and place them in an overall basic protocol (i.e. a procedure) for gene transfer into bacteria. This summary, below, and Figure 4.2 should be studied in conjunction and *carefully*. Figure 4.2 is a key one and will be referred to frequently throughout this chapter and the next two.

- Enzymes were discovered that are capable of cutting DNA at specific sites along its length. These enzymes, called *restriction enzymes*, can be used to cleave long DNA molecules into smaller (gene-size) pieces (Figure 4.2d). 'Gene-size piece' or 'gene-size fragment' are vague terms but, in this context, what is implied generally are pieces/fragments of DNA around 1 000 to 5 000 base pairs in length (see Section 4.3, later). That is, each piece is large enough to comprise just one or a few genes.
- Restriction enzymes can also be used to cause a *single* cleavage in plasmid circles, thus opening a circle, converting a closed circle to a linear molecule of DNA (Figure 4.2b—*note* that although now linear it is convenient to draw such opened circles in a way indicating that they *were* circular).
- Ways were developed of inserting a small (gene-size) piece of DNA into an opened plasmid circle and then closing the circle, effectively 'splicing' the extra DNA into place (Figure 4.2e). This **splicing** is 'seamless' and actually involves covalently integrating the piece of 'foreign' DNA into the polydeoxyribonucleotide chains of the plasmid DNA.
- 'Resealed' ('reclosed') circular plasmids containing foreign DNA are known as **recombinant plasmids**. In the presence of calcium ions such recombinant plasmids can be taken up by other, plasmid-free, bacteria (Figure 4.2f).
- Once inside a bacterial cell a recombinant plasmid can replicate, in so doing copying both its plasmid genes *and* any newly acquired integrated foreign ones (Figure 4.2g); the replicates can replicate, and so on. The bacterial cell can reproduce asexually, as normal (Figure 4.2h); this giving a whole population of genetically identical individuals (called a **clone**) in which each cell contains copies of the same recombinant plasmid.

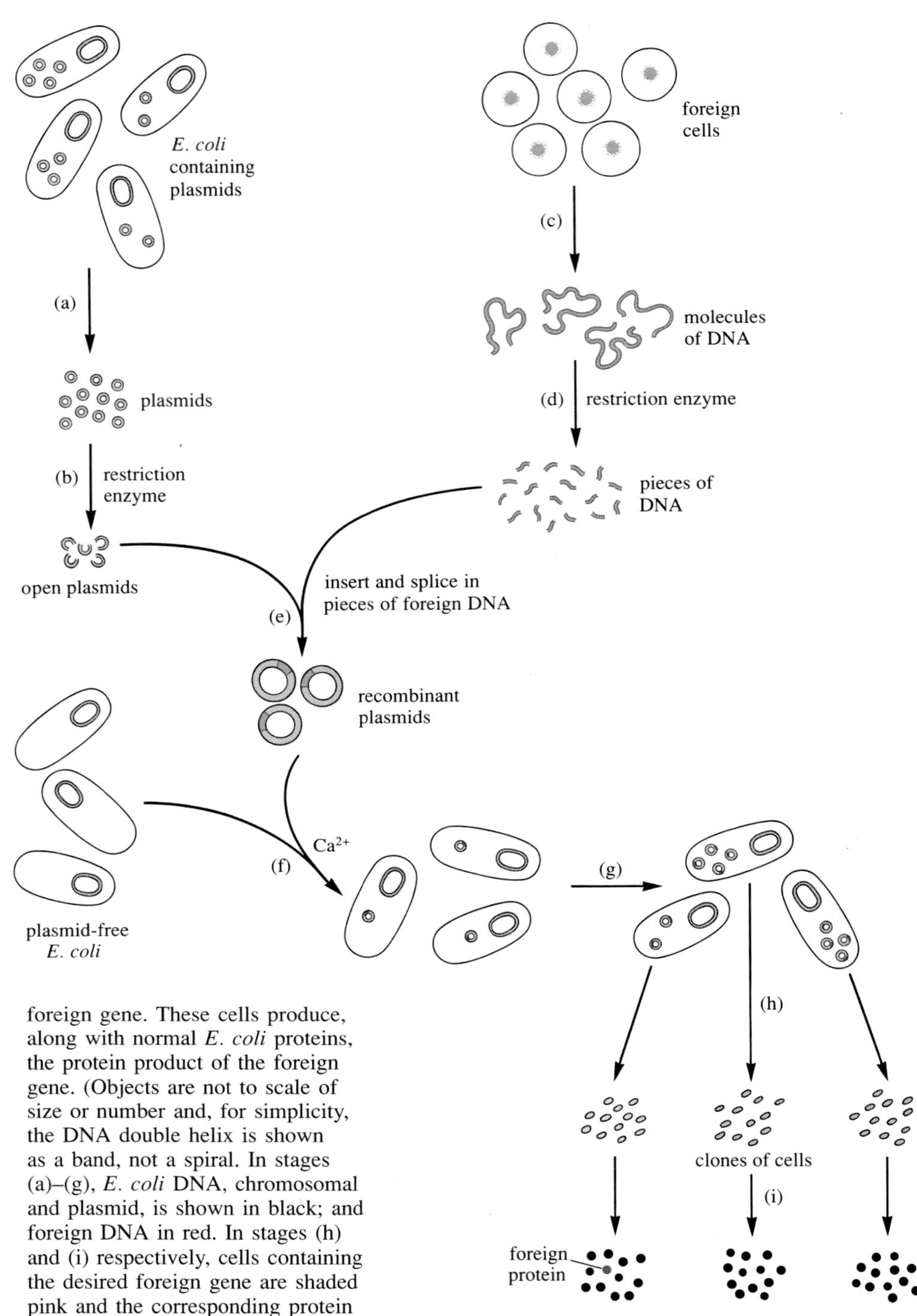

Figure 4.2 An outline protocol for genetic engineering in *E. coli*. (a) Plasmids are isolated from a pure culture of *E. coli*; that is, all the cells derive from a single ancestor. So all the plasmids isolated are identical. (b) The plasmid circles are treated with a restriction enzyme which cleaves each plasmid at the *same* single site thus opening up each circle in the same way. (c) DNA is extracted from some (foreign) cells and the molecules of DNA cleaved (d) into (gene-size) pieces with a restriction enzyme. This results in many *different* gene-size pieces of DNA. (e) Gene-size pieces of the foreign DNA are spliced into the opened plasmid circles which are re-sealed to create recombinant plasmids. Many *different* recombinant plasmids are thus created, because the gene-size pieces of DNA were different. (f) When mixed with some plasmid-free *E. coli*, in the presence of calcium ions, individual recombinant plasmids are taken up by individual cells—*one plasmid per cell*. (g) In each cell containing a recombinant plasmid, that plasmid replicates, replicates can then replicate, and so on—thus the plasmid multiplies. (h) Each cell grows, its DNA replicates (chromosomal *and* plasmid), and the cell divides; its offspring grow, replicate their DNA, divide, and so on. So at each cell division, a cell passes on a copy of its chromosomal DNA *and* identical copies *of its own particular recombinant plasmid* to its descendent cells. Thus *each* cell resulting from step (f) produces a clone of (identical) cells. The *clones* differ from each other according to what particular recombinant plasmid the cells carry; but *within* a clone, *all* the cells contains copies of the same plasmid. Among these clones is one (or perhaps more, see Section 6.1, later) where the recombinant plasmid contains the desired foreign gene. (i) The clones of cells produce proteins. There is a clone (or clones) where the cells contain the desired foreign gene. These cells produce, along with normal *E. coli* proteins, the protein product of the foreign gene. (Objects are not to scale of size or number and, for simplicity, the DNA double helix is shown as a band, not a spiral. In stages (a)–(g), *E. coli* DNA, chromosomal and plasmid, is shown in black; and foreign DNA in red. In stages (h) and (i) respectively, cells containing the desired foreign gene are shaded pink and the corresponding protein product is in red.)

As summarized on p. 33 and in Figure 4.2, following such developments, not only can plasmid genes (say, for antibiotic resistance) be transferred from one strain of *E. coli* to another. More importantly, foreign, non-plasmid DNA can also be transferred into *E. coli*. In effect plasmids can be made to act as carriers, so-called **vectors**, for transferring genes into bacteria. Most significantly, the *foreign DNA transferred can* *be from virtually any source*—from bacteria of the same species as the 'host' ones (i.e. the bacteria receiving the recombinant plasmids; *E. coli* in our example), other species of bacteria, non-bacterial microbes, or even animals or plants.

In essence, that sums up how a foreign gene can be transferred into *E. coli*. Sometimes the aim of the transfer is to get a ready source of the gene itself. But, as will become evident (from Chapter 9, in particular), in many cases it may be not so much the gene that we are after, as its protein product. That is, we may want the specific protein that the foreign gene codes for (Box 5.1). Human insulin and thaumatin are two such protein products already mentioned (Section 3.4). In such cases getting gene transfer is not enough. We must ensure also that, following transfer, the foreign gene is *expressed* in its new surroundings. **Expression** is the term commonly applied to the 'reading' or 'acting out' of the information in a gene to give its protein product (as explained further in Box 5.1, later). So we have included an additional step in our generalized protocol, in Figure 4.2i. All the clones of cells synthesize proteins. Among the many proteins that it synthesizes, *the clone bearing the desired foreign gene* synthesizes the protein coded for by that gene—the protein we are after.

Procedures of the general type outlined in Figure 4.2 have become variously known as **recombinant DNA technology** (frequently abbreviated to **rDNA technology**), or **gene splicing**, or **gene cloning**. The first two names are fairly obvious. The last derives from the fact that the penultimate step (Figure 4.2h) involves **cloning** cells, that is *each* of the cells receiving a plasmid is allowed to grow and divide to give a *separate* population of identical cells (a clone). Each cell in the *same* clone contains the same particular piece of foreign DNA (gene(s)). So, in effect, these foreign genes are also being cloned along with the cells that contain them (Figure 4.2h)—hence 'gene cloning'. The more general term, *genetic engineering*, is also often used for technology like that shown in Figure 4.2, though it can also be taken to cover a wider range of processes.

For any particular clone, the 'actual gene transfer', the uptake of the foreign gene, occurred to its ancestor single cell (at stage (f) in Figure 4.2). But, *in effect*, the same foreign gene has been transferred into the whole clone. The net result is the genetic engineering of a whole clone of cells. We can say that the same foreign gene has been transferred into each of the cells in the same clone.

Of course the protocol outlined above (Figure 4.2) is very sketchy. Also, it is hypothetical but derived essentially from a generalized procedure of a type applicable to *E. coli*. Its main features would, in principle, however apply to many other microbes and, as will be seen from Chapter 8, in many respects to plants and animals too.

However, before we proceed, take another, close, look at Figure 4.2. And remember the whole point of rDNA technology—that triumph of genetic engineering over traditional breeding techniques which we have trumpeted so loudly so far. It is to allow the transfer of *specific* genes from one organism into another (here bacteria). So we seem to be left with an unresolved problem. The procedure shown in Figure 4.2 allows the transfer of *gene-size pieces* of foreign DNA into bacteria but how does it permit the transfer of *specific desired* genes? In other words how does one get a recombinant plasmid, and thereby a bacterial clone from it, *that contains the desired gene*? Does one pick out the desired gene from the cleaved foreign DNA, or what?

We shall now consider the main features of the scheme in Figure 4.2 in more detail: cleaving the foreign DNA into gene-size pieces; splicing the foreign DNA into the plasmids; getting cells to take up the recombinant plasmids; then, the central point of the whole procedure—getting the clone with the *desired* gene; and, lastly, often equally important, getting its protein product. To do all this will carry us right through to the end of Chapter 6. First then, how to cut the foreign DNA into gene-size pieces (Section 4.3, on p. 38).

4.2

Box 4.2 The structure of DNA

DNA is a polymer whose full name, **deoxyribonucleic acid**, reveals that the units (**monomers**) of this polymer are **deoxyribonucleotides**. Deoxyribonucleotide is connected to deoxyribonucleotide via covalent linkage involving a phosphate group, giving a **polydeoxyribonucleotide** chain. A polydeoxyribonucleotide chain is shown schematically in Figure 4.3.

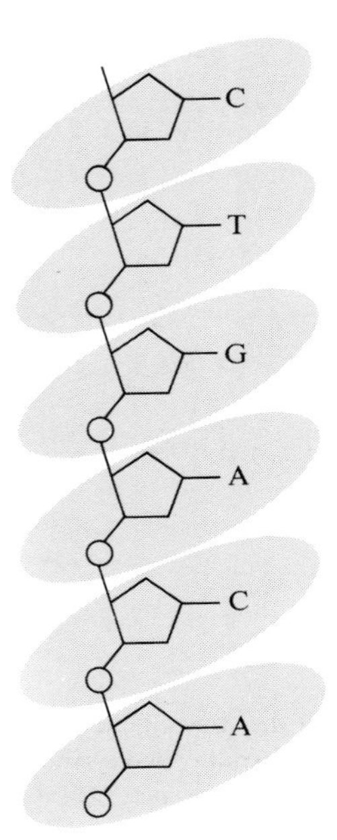

Figure 4.3 Schematic diagram of a portion of a long polydeoxyribonucleotide chain. Each shaded oval is one deoxyribonucleotide; deoxyribonucleotides differ according to the bases that they contain. The four types of base are shown as A, T, C and G. Each deoxyribonucleotide is connected to the next in the polydeoxyribonucleotide chain by covalent links involving a phosphate group (phosphate groups are shown here as circles). This backbone of the chain consists of alternating phosphate groups and deoxyribose sugar components (shown as pentagons). The sequence of deoxyribonucleotides (i.e. the sequence of bases) shown has no particular significance; we could have chosen any sequence.

From Figure 4.3 you will note that there are four types of deoxyribonucleotide. They differ according to which base they contain. There are four types of base: two, **adenine** (**A**) and **guanine** (**G**) are, chemically, **purines** (**Pu**); the other two, **cytosine** (**C**) and **thymine** (**T**) are **pyrimidines** (**Py**). As a whole each deoxyribonucleotide comprises three covalently linked parts—**deoxyribose**, a phosphate group, and one of the four bases.

In principle, the different deoxyribonucleotides (i.e. differing in their bases) can occur in any sequence along the polydeoxynucleotide chain and it is these base sequences that 'spell out', encode, the information in the genes. The genetic information in DNA resides in the sequence of bases (Box 4.1).

But containing information is only one aspect of DNA. The other is the ability to pass this on generation after generation—the information must be capable of being copied, being replicated. Faithful replication of DNA depends on another key property of this molecule. DNA is not a single long polydeoxyribonucleotide chain but in fact comprises

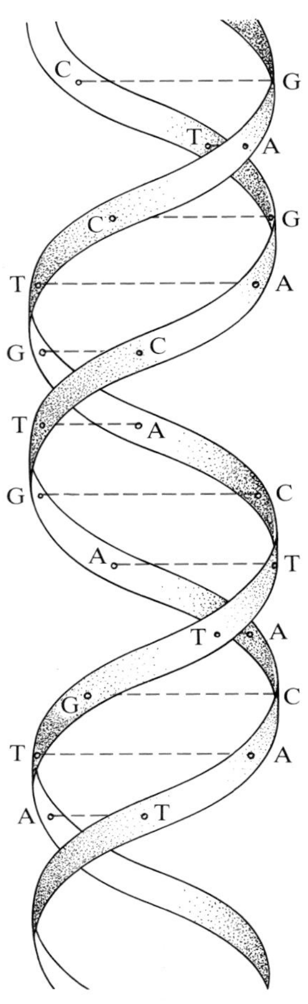

Figure 4.4 Schematic diagram of the DNA double helix. The spiralling backbone of each strand, shown as a broad ribbon, is composed of alternating deoxyribose and phosphate with paired bases connecting the two strands. The bonds holding these paired bases, and hence the two strands, together are hydrogen bonds (indicated simply as *single* broken red lines).

two different chains, two strands, wound round each other to form a double helix, a structure represented in Figure 4.4.

Critically, unlike the strong covalent bonds involved in linking deoxyribonucleotide to deoxyribonucleotide in each of the two polydeoxyribonucleotide strands, the bonds holding the two strands together in the double helix are weak bonds. These are **hydrogen bonds** which form between pairs of bases opposite each other in the double helix, that is bases that lie one in each of the two polydeoxyribonucleotide strands. Viewed as a double helix, a DNA molecule can be regarded as containing *base pairs*. (Thus though within any one gene just *one strand* of the DNA contains the information, we still sometimes speak of a gene being 'so many base *pairs*' long, physically; see also Box 4.1.) Most importantly, there are **base-pairing rules**, certain bases must pair up specifically with other bases — adenine (A) with thymine (T), and cytosine (C) with guanine (G). These base pairs, **A–T** and **G–C** (and **T–A** and **C–G**, as the pairing rules hold in both directions) are known as **complementary base pairs**. It is base complementarity that ensures faithful replication of the DNA.

We shall not go into the details of DNA replication. But in essence, the two strands of the double helix separate — hence the importance of the *weak* hydrogen bonds holding the two strands together. Each strand can then act as sort of 'die' or 'printer's template' for building up a partner (complementary) strand to itself. This complementarity is inherent in the base complementarity, according to base-pairing rules; that is, wherever there is an exposed (i.e. unpaired) adenine in a separated strand it will pair up with a thymine, a guanine with a cytosine, a thymine with an adenine, and a cytosine with a guanine. In the cell there are free deoxyribonucleotides that can pair up with exposed bases in the separated strands. Having been thus 'attracted' to the separated strands, these deoxyribonucleotides can be joined together covalently to form new polydeoxyribonucleotide strands. The net result is two identical double helices from what was one; in each of the helices one strand is from the old helix (*conserved*), the other is newly formed. This replication is thus said to be **semi-conservative**. It is shown very simply and schematically in Figure 4.5.

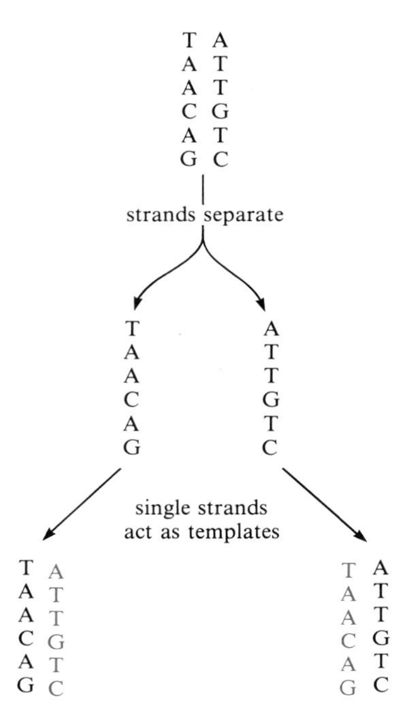

Figure 4.5 Semi-conservative replication of DNA. Each strand of the double helix is shown simply by the base sequence of that strand; the features identical in each strand (the backbones of deoxyribose–phosphate), the spiralling, and the hydrogen bonds between base pairs being ignored. New strands in the newly replicated DNA molecules are shown in red, the old (conserved) ones in black.

Finally note that in this book generally we shall drop the cumbersome 'deoxyribo' simply referring to **nucleotides** (or just bases) in DNA. ■

To test your understanding of the material in Box 4.2 you should attempt Questions 4.2 and 4.3 below.

Question 4.2 Given below is the sequence of bases in one strand of a DNA double helix. What is the sequence in the other strand?

AACGGGTTACCTGTACCAAAT

Question 4.3 Which of the following statements are true and which are false?

(a) The sugar found in DNA is ribose.

(b) Each nucleotide (more correctly *deoxyribo*nucleotide) in DNA comprises a sugar, a base and a phosphate group covalently linked to each other.

(c) Base-pairing rules ensure that adenine pairs with cytosine and thymine with guanine.

(d) DNA would be a more efficient molecule (i.e. better adapted to its evolutionary function) if there were covalent bonds holding together the base pairs between the two strands of the double helix.

4.3 Chopping up foreign DNA

So, naturalists observe, a flea
Hath smaller fleas that on him prey;
And these have smaller still to bite 'em;
And so proceed ad infinitum.

(Johnathan Swift, *On Poetry: a Rhapsody*, 1733)

We are only too familiar with the range of human disease caused by bacteria and viruses. One of our greatest protections against infectious disease is our *immune system* of which a major property is that it can distinguish *self* from *non-self*. Thus invading bacteria or viruses can be recognized as 'foreign' and an array of defences be brought into action to help eliminate any threat that they present. But though unicellular and hence tiny, bacteria themselves *suffer* 'disease' from yet tinier viruses. (The discovery of the Lilliputian world of viruses actually occurred nearly 200 years after Swift's day.) Viruses are generally *host-specific*, that is, each type of virus is able to infect one or just a few species of organism and not others. As a whole, viruses that infect bacteria are usually called **bacteriophages** or **phages**. Like us, bacteria are not totally unprotected against viral (phage) infection.

It appears that bacteria also possess a sort of 'immune system'. In detail it operates quite unlike our own but does share the key property of distinguishing self from non-self. It recognizes non-self DNA; as might belong to an invading phage, for example. The recognition is achieved by specific enzymes, as is the response—cleaving the non-self DNA into smaller pieces. As viruses can only reproduce inside host cells and need their viral genes (DNA) intact to do so, chopping up their DNA is an effective bacterial defence against phages. (Incidentally, there are certain viruses, including some phages, that have *ribonucleic acid*, *RNA*, as their genetic material and where DNA does not feature in their life cycles. We are, however, not concerned with such RNA phages here.)

In considering how this bacterial 'immune system' works it is necessary to focus on the enzymes that recognize and cleave foreign (non-self) DNA. As such enzymes probably cleave the DNA of invading phages and hence *restrict* phage growth, they are known collectively as **restriction enzymes**. The key question is thus: how does such a restriction enzyme distinguish between self and non-self DNA and hence cleave only the latter? We can put this another way and present it as two separate questions:

1 How does a restriction enzyme recognize, and hence cleave, non-self DNA?

2 Why does it not cleave its own (self) DNA?

We shall take these questions in order.

▷ How, in principle, do you think recognition of a particular DNA molecule might be achieved? That is, how could a restriction enzyme distinguish one molecule of DNA from another (say, self from non-self)?

▶ As you are aware, a key property of all enzymes is their *specificity* for their substrates. So part of the answer lies undoubtedly in a restriction enzyme's recognition of DNA as its substrate for cleavage. But it does not recognize all DNA as appropriate for cleavage, just non-self DNA. So we must consider what unique feature any particular type of DNA molecule might possess. The most obvious is a unique sequence of (deoxyribo)nucleotide bases—that is, DNA from one source will differ in its base sequence from DNA from another source.

It would seem unlikely that a restriction enzyme needs the complete base sequence of a DNA molecule to distinguish it from another DNA molecule (the molecules may well be thousands or millions of bases long!). In fact each type of restriction enzyme recognizes a very short specific sequence of bases—its so-called **target sequence**—typically some four to seven base pairs long. Table 4.1 gives a small selection of restriction enzymes showing the bacteria they occur in and their target sequences.

Table 4.1 The characteristics of some restriction enzymes. (Cleavage points are indicated by vertical arrows.)

Enzyme	Source	Target sequence and cleavage points
Hind II	*Haemophilus influenzae*	G T Py↓Pu A C C A Pu↑Py T G
Hae III	*Haemophilus aegyptius*	G G↓C C C C↑G G
Bal I	*Brevibacterium albidum*	T G G↓C C A A C C↑G G T
Alu I	*Arthrobacter luteus*	A G↓C T T C↑G A
Eco RI	*Escherichia coli*	G↓A A T T C C T T A A↑G
Hpa II	*Haemophilus parainfluenzae*	C↓C G G G G C↑C
Pst I	*Providencia stuartii*	C T G C A↓G G↑A C G T C
Hind III	*Haemophilus influenzae*	A↓A G C T T T T C G A↑A
Bam HI	*Bacillus amyloliquefaciens*	G↓G A T C C C C T A G↑G

Note: Some enzymes (e.g. *Bal* I) cleave both strands of the DNA at points exactly opposite each other, producing flush-ended (blunt-ended) DNA fragments. Other enzymes (e.g. *Eco* RI) make staggered cuts, yielding ragged-ended DNA fragments. Both types of fragment are suitable for cloning. Note also that *Hind* II can 'act on' more than one type of target sequence. (Py means pyrimidine and Pu means purine.) It can act when either cytosine (C) *or* thymine (T) occupy the Py positions, and either adenine (A) *or* guanine (G) the Pu positions.

The cleavage points for each of the enzymes shown are *within* the target sequences as indicated by the arrows. Thus, for example, we can see that the enzyme *Hae* III occurs in *Haemophilus aegyptius* and has

GG↓CC
CC↑GG

as its target sequence, and cleaves the complementary strands of DNA, at the points indicated by the arrows.

Thus, in general, where a non-self DNA molecule contains the target sequence for a particular restriction enzyme at *one or more sites* along the length of the double helix, that restriction enzyme will cleave the molecule at *each* of those sites. So, for example, a DNA molecule with the sequence

GG↓CC
CC↑GG

occurring at three different sites along its length would be cleaved by *Hae* III into four fragments of DNA, i.e. *restriction enzyme fragments* (Figure 4.6).

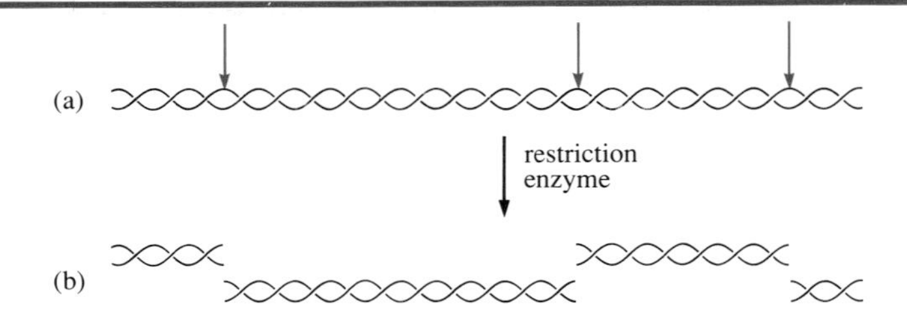

Figure 4.6 Restriction enzyme cleavage of a DNA molecule. The target sequence for a particular restriction enzyme occurs at three sites (arrowed) in the hypothetical DNA molecule shown in (a). On treating this DNA with the enzyme, each site is 'attacked', this generating four restriction enzyme fragments (b) from the single molecule. (Not to scale.)

This explains how a restriction enzyme can recognize non-self DNA and cleave it. But it still leaves our other question: why does it not cleave its own (self) DNA? For example, why does *Hae* III not cleave the indigenous (self) DNA within its own cell, *H. aegyptius*?

It seems that where bacteria possess a restriction enzyme they also possess another, 'corresponding', **modification enzyme**. This can *modify* bases within the cell's own DNA that would be within *potential* target sequences for that particular restriction enzyme. The modification enzyme is usually a *methylase* (i.e. it catalyses the addition of *methyl* ($-CH_3$) *groups*). A methyl group is added to either an adenine (A) or cytosine (C) within potential target sequences. Such modification occurs after DNA is synthesized in a cell. The methyl groups do not affect base pairing in the double helix (e.g. methylated cytosine still pairs with guanine) but do protect it against attack by the cell's restriction enzyme to which that modification enzyme 'corresponds'. Thus, for example, in *H. aegyptius* DNA, what would be target sequences for *Hae* III in foreign DNA have been modified by methylation of one of the cytosines (C ⟶ C*) in each strand:

GGCC ⟶ GG C*C
CCGG   CC*G G

This renders that sequence no longer a target for *Hae* III. We might anticipate that for each type of restriction enzyme inside a bacterium there is likely to be a corresponding modification enzyme.

Thus, this combination of specific restriction enzymes and corresponding modification enzymes helps the cell cleave foreign DNA (non-self and hence not protected by modification) while protecting its own DNA against its own restriction enzyme(s).

So much for restriction enzymes *within* bacteria. For our purposes, the critical fact is that we can isolate and purify restriction enzymes and use them *in vitro*, 'in the test-tube', for rDNA technology. The first such restriction enzyme to be clearly described was *Hin*d II (see Table 4.1). When used *in vitro*, it was found to break down non-self DNA, but not its own (*Haemophilus influenzae*) DNA; consistent with its behaviour in living *H. influenzae*.

Since the discovery of *Hin*d II in 1970, hundreds of restriction enzymes have been isolated from a wide variety of bacteria. Though not all have been shown to be responsible for any specific cellular function (say, bacterial 'immunity' against phages), they all have the property of cleaving *foreign* (i.e. non-self) DNA and not that from their own cellular source—this is indeed how a restriction enzyme can be defined.

Thus in practice, whatever the source of the DNA that one wishes to chop up, there is likely to be a suitable method available. There are three aspects to this suitability:

1 We must of course find an enzyme capable of cleaving the DNA in question; the DNA must contain target sequences compatible with the enzyme's specificity. For

example, it would be no good trying to cleave DNA from T7 (a phage) with the enzyme *Eco* RI (Table 4.1)—somewhat remarkably, although T7 DNA is 40 000 base pairs long, the sequence

GAATTC
CTTAAG

does not occur at all within it, so *Eco* RI will not cleave it.

2 The ultimate aim of the whole procedure shown in Figure 4.2 is to transfer one or more *specific* desired foreign genes into bacteria. You will recall that the purpose of cutting up the DNA is to reduce it to gene-size pieces (Figure 4.2d) so that each such piece can be inserted into a separate plasmid (Figure 4.2e). Somewhere among such pieces we hope is one within which is our desired gene (*a gene which we require intact*). That is, we need our entire gene (its whole base sequence) to be part of the base sequence of a *single* gene-size fragment of DNA. It is no use having the gene itself fragmented, bits of it within different gene-size pieces of DNA. Therefore the restriction enzyme must *not* cleave within the desired gene. Genes are fairly long sequences of bases (often 1 000 base pairs or more; but see Box 5.2, later) and an enzyme making very frequent cuts in the DNA *might also cut the desired gene into pieces*. For example, the enzyme *Alu* I (Table 4.1) recognizes and cleaves the base sequence

AGCT
TCGA

This sequence occurs about 5 000 times in the single molecule of DNA that is the *E. coli* chromosome (Figure 4.1). As this molecule is only about 4 million base pairs long, cuts would occur *on average* every 800 base pairs (this ignores any possible non-random occurrence of the sequence recognized by *Alu* I). Thus if in this instance *E. coli* DNA were our *foreign* DNA (i.e. for transfer into some other organism), using *Alu* I would run a high risk of cuts occurring *within* a gene desired for transfer. So, using this technique, to minimize the chances of cleaving the required gene it would be advisable to find a restriction enzyme that does not have its target sequence too frequently represented in the particular foreign DNA and hence does *not* cleave that DNA into too small pieces.

3 On the other hand, the restriction enzyme used must *not* yield pieces of DNA that are too large. Many plasmids can cope with DNA up to a few thousand base pairs long but larger pieces of DNA present problems. Recombinant plasmids (Figure 4.2e) do not readily tolerate large pieces of inserted DNA. Large inserts tend to be lost during replication of the plasmids—so we would risk losing the very thing we want, the foreign gene.

As I have already implied, the outline given in Figure 4.2 is a generalized one—the actual protocols used vary round this schematic theme. For example, there are variants (which I need not detail) of DNA cleavage using restriction enzymes (Figure 4.2d), that both yield relatively large fragments and improve considerably the chances of getting a desired foreign gene intact. It is evident that this increases our chances of finding a clone with the desired foreign gene (Figure 4.2h). (Producing larger fragments does mean, however, that plasmid vectors may not be always the most suitable for cloning; Section 7.1.)

However, we rarely know the base sequence of a desired gene in advance and almost certainly do not know the base sequence of the entire foreign DNA. So, whatever the technique, for any particular gene transfer, there may well be a certain amount of trial and error.

Activity 4.1 *You should spend up to 15 minutes on this activity.*

Let us assume for the sake of this particular activity, that we do know something of the base sequence of a desired gene and its surrounding DNA.

Given in Figure 4.7 is a tiny part of the base sequence of this hypothetical foreign DNA molecule; the region shown in red is a gene, some 1 200 base pairs long, that we wish to transfer into *E. coli*. Where the base sequence is unknown it is given as a sequence of three dots, though the length of such regions is not to be taken as indicated by the number of dots.

Obviously you cannot make firm predictions about unknown sequences. But, based on what is known, which of the following restriction enzymes would be suitable for cleaving the DNA in Figure 4.7 so that the required gene is within a small (gene-size) fragment to be then spliced into a plasmid, and why?

Bal I, *Alu* I, *Hae* III and *Eco* RI

You will need to refer to Table 4.1. To save flipping between this page and the table it might be helpful to copy down the relevant information from the table for the four enzymes listed above.

...TTCAGCTAGGCCCTATGCTAGCTACC... AGTTCCTAATAAGAGGCCTAAGCTTATCCTA...

...AAGTCGATCCGGGATACGATCGATGG... TCAAGGATTATTCTCCGGATTCGAATAGGAT...

Figure 4.7 For use with Activity 4.1.

4.4 Splicing in the genes

Having got suitable gene-size pieces of the foreign DNA (neither too small nor too large) we now need to splice these into plasmids to create recombinant plasmids (Figure 4.2e). In essence, this is a four-stage procedure:

1 A number of identical plasmids (vector plasmids) are isolated from some *E. coli* (Figure 4.2a).

2 Each plasmid circle is opened (Figure 4.2b).

3 A gene-size piece of foreign DNA is inserted into each opened plasmid.

4 The foreign DNA must be sealed/spliced in, leaving each plasmid as a closed, recombinant, plasmid circle (Figure 4.2e).

▷ How might we open a plasmid circle?

▶ Use a restriction enzyme.

Thus, opening a plasmid circle (Figure 4.2b) is essentially the same as cleaving the linear molecules of foreign DNA (Figure 4.2d). *But* the type of plasmid used and the enzyme used to cleave it must be so chosen that *only a single cleavage occurs*. In this way each circle is transformed into a linear (double-helical) DNA molecule, *not* broken into two or more fragments of DNA.

Thus, for example, if one wishes to so linearize the plasmid called pSC101 (one used in *E. coli*), *Eco* RI is a suitable enzyme, there being only one target sequence for this enzyme in the plasmid.

▷ What is the target sequence for *Eco* RI? (See Table 4.1.)

▸ As shown in Table 4.1 its target sequence is

GAATTC
CTTAAG

The next stage is to get gene-size pieces of foreign DNA inserted into the opened (linearized) plasmid circles. There are a number of techniques for doing this and the ones used depend on the circumstances. Here we need not consider any in detail. Suffice it to say that, using such techniques, when mixed together gene-size pieces of foreign DNA and opened plasmid circles will associate. A foreign DNA fragment inserts into the gap in an opened plasmid circle but this insertion involves *weak binding*. This unstable relationship between foreign DNA fragments and opened plasmid circles can be cemented permanently using an enzyme called *DNA ligase* (more often just called **ligase**). This enzyme catalyses the formation of covalent deoxyribose–phosphate bonds, the bonds linking nucleotide to nucleotide in each strand of a DNA double helix (Box 4.2). So the associated ends of the two types of DNA (plasmid and foreign fragment) are sealed, spliced into the same two (double helix) polydeoxyribonucleotide strands. The result is a single double helix of DNA comprising both plasmid DNA and foreign DNA, the double helix being a complete covalently closed circle—a *recombinant plasmid* (Figure 4.2e).

4.5 Getting the recombinant plasmids into bacteria

Having created recombinant plasmids it now remains to get those plasmids taken up into the host, *E. coli*, cells (Figure 4.2f). This is achieved by mixing *E. coli* that have no plasmids with recombinant plasmids in the presence of Ca^{2+}. Under such conditions each *E. coli* can stably take up just one plasmid alone. This ensures that only one type of recombinant plasmid is maintained in any particular cell. However, once in a cell, a recombinant plasmid may multiply to produce several copies of that particular plasmid per cell (Figure 4.2g). When each of these cells is then allowed *separately* to grow and divide, all of its descendent cells will contain copies of just the one particular type of recombinant plasmid. Each cell will thus produce a clone of descendants; all cells within the same clone will contain the same type of recombinant plasmid. (Put another way, the foreign gene(s) may differ between different recombinant plasmids. But within each clone, every cell carries the same type of recombinant plasmid and hence the same foreign gene(s).) The cloning of foreign genes will be complete (Figure 4.2h)—each clone of bacteria will contain its own particular foreign gene(s).

However, there is one further consideration about the particular *strain* of *E. coli* used as a host. It may be advisable to use a mutant strain that lacks the natural *E. coli* restriction enzymes.

▷ Why should this be advisable?

▸ We wish to transfer a recombinant plasmid containing a foreign gene into *E. coli* and get it to remain stably and replicate there. Normal **wild-type** *E. coli*, the strain isolated from the wild, will contain its own restriction enzymes, such as *Eco* RI (Table 4.1). These *might* attack and cleave the incoming recombinant plasmid DNA (as foreign DNA) and hence destroy the very thing we wish to propagate in those *E. coli* cells. Using a strain that lacks such restriction enzymes avoids such events.

Two problems remain. One is to get the desired foreign gene *expressed*—that is, producing its protein product (Figure 4.2i). The second returns to the unresolved problem mentioned in Section 4.2 about the protocol shown in Figure 4.2. There are very many different clones, each clone with its own particular recombinant plasmid in the cells and hence its own particular foreign gene(s). How do we know which of the many clones actually contain the *desired* foreign gene and hence will make the *desired* foreign protein? Chapters 5 and 6 tackle these problems.

Summary of Chapter 4

1 In prokaryotes the genetic material, DNA, is not bound up with proteins to form eukaryote-like chromosomes. The *DNA is essentially 'naked'*.

2 In *E. coli* the chromosome is a closed circle of naked DNA double helix. In addition to this single chromosome, *E. coli* can contain one or more small circles of DNA called *plasmids*.

3 In general only one type of plasmid can exist inside a single bacterium.

4 Plasmids can serve as carriers, *vectors*, of foreign DNA into bacteria.

5 Typically, as in *E. coli*, transferring foreign DNA (genes) into bacteria involves a number of key steps: *cutting* the foreign DNA into *gene-size pieces; inserting and splicing* the foreign DNA fragments into opened plasmid circles to form *recombinant plasmids; uptake* of the recombinant plasmids by bacteria that have no plasmids.

6 If the *protein product of the foreign gene* is desired, then the foreign gene once transferred into bacteria must also be *expressed.*

7 Cutting, cleaving, the foreign DNA into gene-size fragments can be achieved using *restriction enzymes*. Each type of restriction enzyme has a specific *target sequence* in DNA that it can cleave.

8 Restriction enzymes *do not cleave DNA from their own source* (i.e. the same bacteria). Bases in potential target sequences in such DNA have been *modified*, rendering these sequences resistant to the corresponding restriction enzyme.

9 Restriction enzymes that have a single target sequence in plasmids can be used to *open*, to *linearize*, the plasmid circles.

10 *Inserting* foreign DNA fragments into opened plasmid circles can be achieved in several ways.

11 Once inserted in an opened plasmid circle, foreign DNA can be sealed in using *ligase* to create a recombinant plasmid.

12 Recombinant plasmids can be taken up by bacteria in the presence of *calcium ions* (Ca^{2+}).

13 The overall process of gene transfer into bacteria is known variously as *recombinant DNA (rDNA) technology*, or *gene splicing* or *gene cloning*. Sometimes it is known simply as genetic engineering but this also has a wider meaning.

Question 4.4 For each of items (a)–(c) below, which of the following are true: (i) contains DNA; (ii) contains histones; (iii) occurs in the cell nucleus?

(a) the *E. coli* chromosome

(b) bacterial plasmids

(c) a human chromosome.

Question 4.5 Assuming you are considering a procedure for transferring foreign genes into bacteria, put the following steps in sequence:

(a) cleave the foreign DNA;

(b) put the bacteria plus recombinant plasmids in a medium containing calcium ions;

(c) grow a clone from each bacterium;

(d) insert and splice the foreign DNA fragments into opened plasmid circles.

Question 4.6 Look at (a)–(d) in Question 4.5. In which steps might the following be needed and under what circumstances?

(i) ligase

(ii) restriction enzyme.

5 Getting the right gene product

This chapter continues the tale begun in Chapter 4—how to transfer genes into bacteria so that they produce their protein products. So keep Figure 4.2 handy. Once again there is some technical detail which will reward close study though it is the principles that you should understand and be able to recall. Though we give a summary at the end of the chapter, preparing your own may again prove useful.

By the end of Chapter 4 we had achieved most of the steps in Figure 4.2 (a–h); ending with clones of bacteria, where each clone carries its own specific foreign gene. (Put more precisely, the foreign gene can differ between different clones but in each clone each cell carries the same specific foreign gene—see also Section 6.1, later.) It now remains to ensure that the protein products of the foreign genes can be expressed in their new bacterial surroundings (Figure 4.2i) and that we can spot the clone that carries the particular foreign gene, and its protein product, that we want.

5.1 Expressing genes

As you know, the decoding of the information in a gene to produce its protein product involves essentially two phases—*transcription* and *translation* (these processes are detailed in Box 5.1). The first results in the production of a transcript, a copy, of the gene in the form of a complementary messenger RNA (mRNA). The second phase requires the translation, the decoding, of the message to produce a polypeptide chain that has the correct amino acid sequence. The code itself, the genetic code, is essentially universal; that is, it is virtually the same in all living organisms—a fact useful for genetic engineering.

▷ Why is the universal nature of the genetic code useful for genetic engineering?

▶ One of the main objectives of genetic engineering is to transfer genes between different species of organism *and* get the protein products of those genes produced in their novel surroundings. Indeed, it may be a protein product, *as such*, that we are after, say, thaumatin. Or, it may be that we do not want the protein product, as such, but do need its presence in the cell so as to change the cell's metabolism in some particular way (producing a starch-degrading enzyme in *S. cerevisae*, Section 3.4, would be one such example). In either case we need the cell to produce the protein product of a foreign gene transferred into it. Say that the genetic code were not universal and different species used different codes for the same process. Then, though we could transfer genes between species, those genes could not be accurately decoded in their new surroundings to give their normal products, that is, the same products as in their original species. For example, the gene for thaumatin, a plant protein, in bacteria might not be expressed at all or might be expressed to give an abnormal product—it might be like trying to translate from French to English using a French–Spanish dictionary. The essentially universal nature of the genetic code means that in principle any gene can be decoded to give the same product in any organism.

5.1

Box 5.1 DNA, mRNA and protein

In some senses the total DNA of a cell is like an instruction manual. The 'instructions' are for making cellular components and each gene can be regarded as a 'piece of information', one such instruction. The 'product' or 'output' of acting out the instruction in a particular structural gene (Box 4.1) is a specific polypeptide chain. The process of acting out the instruction in a gene is termed gene expression.

But not all the instructions (genes) in a cell's DNA need be acted upon (expressed) in that cell, nor are all instructions acted upon at the same time. Different genes are 'switched' on or off at particular times. And in multicellular organisms, different genes may be switched on in different types of cell, though all the somatic cells contain the same DNA. The switching mechanism itself involves instructions contained in the DNA and certain genes (control genes) are involved in switching some structural genes on and off (Box 4.1).

Of course an instruction manual for constructing a car, a house or whatever cannot of itself do the building. Something or someone must read, interpret, the instructions; use the appropriate materials and tools; do the building work. Likewise, there must be a system in cells for interpreting and acting upon (expressing) the instructions in the DNA, the genes. Each structural gene, when switched on, must be 'read', its instruction acted upon to produce the correct polypeptide product. But before considering this process, it is worth recapitulating briefly what the ultimate product of gene expression, a **polypeptide chain** (often just called a **polypeptide**), is actually like.

Chemically, all proteins are composed of one or more polypeptide chains. As the name suggests, a polypeptide is a polymer. Polypeptide chains can be many hundreds or even thousands of units long. The individual units, the building blocks or monomers, in a polypeptide are **amino acids**.

All amino acids can be represented by the chemical formula shown below, where the so-called **R group** can vary:

$$H_2N-\underset{\displaystyle R}{\underset{|}{CH}}-COOH$$

In polypeptides 20 different types of amino acid occur; they differ from each other in that their R groups differ. In a polypeptide chain, adjacent amino acids are joined together via **peptide bonds** that form between the **carboxyl group** (—COOH) of one amino acid and the **amino group** (—NH_2) of the next one. This results in a chain with a free amino group at one end and a free carboxyl group at the other end. By convention the chain is written with the (amino) **N-terminal residue** at the left end and the (carboxyl) **C-terminal residue** at the right. (When combined in a polypeptide chain each amino acid is called more properly an **amino acid residue**.) We can represent a polypeptide as shown at the bottom of the page.

We can now see that gene expression 'reads' the information in a structural gene, in DNA, and uses this to determine the particular sequence of amino acids in the corresponding polypeptide. This expression of a structural gene occurs in two phases.

The first phase is known as **transcription**. The word itself is very informative, something not universally true in science. In everyday meaning, 'transcription' is a process whereby one form of communication, speech, is copied into another, written or typed words. The physical form of the communication changes (from sound waves to ink on paper) but *the language remains the same*. Similarly, in transcribing a gene the 'language' is unchanged but the copy is in a somewhat different physical form—RNA not DNA.

▷ What is the 'language' of DNA?

▶ The language, the instructions in DNA, are 'written' (encoded) as base sequences. In each gene the instruction is a specific base sequence in just one of the two strands (the *coding strand*) of the DNA double helix (Boxes 4.1 and 4.2).

The language in RNA is the same. In RNA too the instructions are encoded as base sequence, the four bases in RNA being the same as those in DNA except for thymine which is replaced by **uracil (U)**. Critically, the pairing rule for uracil is the same as that for thymine, it pairs with adenine; the other bases in RNA, cytosine and guanine are the other complementary pair (Box 4.2). But unlike DNA, RNA is a *single-*

$$H_2N-\underset{R_1}{\underset{|}{CH}}-\underset{O}{\underset{\|}{C}}-\underbrace{NH-\underset{R_2}{\underset{|}{CH}}-\underset{O}{\underset{\|}{C}}}_{\text{one amino acid residue}}-\cdots-NH-\underset{R_n}{\underset{|}{CH}}-COOH$$

Structure of a polypeptide. (The dots denote the presence of many more amino acid residues.)

stranded (single chain) molecule, a **poly-ribonucleotide** to give it its full chemical name (where the sugar in each (ribo)nucleotide is ribose as against deoxyribose in DNA). (As with (deoxyribo)nucleotides, Box 4.2, we can simply call ribonucleotides *nucleotides*). RNA is formed by copying the base sequence in one strand, the coding strand, of a stretch of DNA (a gene) according to the usual base-pairing rules—where A occurs in the DNA strand, U is inserted in the forming RNA; where C is in the DNA, a G is inserted in the RNA; and so on. An enzyme, **RNA polymerase**, catalyses the joining together of the incoming ribonucleotides. The formation of an RNA strand by transcription from DNA is shown schematically in Figure 5.1.

The coding strand of the DNA here acts as a **template** for forming the RNA. Of course the RNA copy contains a base sequence *not* identical to the coding strand of the DNA but *complementary* to it. (The RNA base sequence is thus identical to that in the *non-coding* strand of the gene, except U replaces T, as a glance at Figure 5.1 will show. Any-how, the net result is a faithful transcript of one of the two strands of the portion of DNA that comprises the gene in question.) Such an RNA transcript of a structural gene is called, appropriately, a **messenger RNA** or **mRNA** for short. Appropriate as it carries the instruction, the message, from the DNA to the *ribosomes* where this message is *translated* to give polypeptide.

Translation is the second of the two phases of carrying out the instruction in a structural gene. Once again, the term is informative. As implied, translation involves a change in language. The 'four-letter language' of base sequence (as in DNA and mRNA) is translated into the '20-letter language' of polypeptides, each of the 20 'letters' being a different amino acid. Going from a four-letter language to a 20-letter one requires *groups* of bases to be used to encode the 20 possible amino acids (just as to express English in the two-letter language of Morse, we need groups of dots and dashes). To code for 20 different amino acids would *in theory* require groups of *at least* three bases in the code (this actually allows 64 different groups of three). And, it is in fact the case that groups of three bases are used in living organisms. A group of three bases, a **triplet**, encodes a single specific amino acid. 64 triplets exist and each specific amino acid is thus encoded by one or more specific **codons**; codon being the term for a triplet of bases. The overall 'dictionary', the **genetic code**, is shown in Table 5.1. Note that the codons shown there are triplets found in the *mRNA* not the DNA. The genetic code is essentially *universal*, that is, it is the same in all living organisms, with one or two very minor exceptions.

The actual process of translation occurs on ribosomes where the mRNA is translated one codon at a time to produce the corresponding polypeptide chain. Another type of RNA,

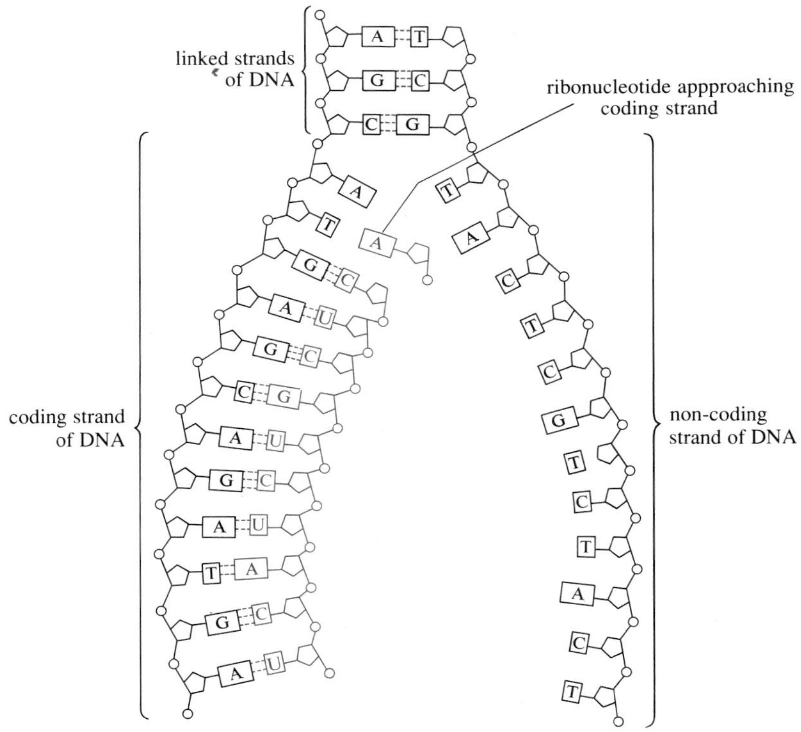

Figure 5.1 Transcription. Where a gene is being transcribed that portion of the DNA double helix is *temporarily* unwound. This permits the sequence of bases in just one strand of the double helix, the coding strand, (i.e. that containing the instruction) to be copied to give a complementary RNA (shown in red). The fidelity of the transcription depends on complementary base pairing. The joining together of the ribonucleotides to form the growing RNA strand is catalysed by an enzyme called RNA polymerase.

Table 5.1 mRNA codons and the amino acids to which they correspond.

First letter	Second letter: U	Second letter: C	Second letter: A	Second letter: G	Third letter
U	UUU, UUC **Phe**; UUA, UUG **Leu**	UCU, UCC, UCA, UCG **Ser**	UAU, UAC **Tyr**; UAA stop; UAG stop	UGU, UGC **Cys**; UGA stop; UGG **Trp**	U, C, A, G
C	CUU, CUC, CUA, CUG **Leu**	CCU, CCC, CCA, CCG **Pro**	CAU, CAC **His**; CAA, CAG **Gln**	CGU, CGC, CGA, CGG **Arg**	U, C, A, G
A	AUU, AUC, AUA **Ileu**; AUG **Met**	ACU, ACC, ACA, ACG **Thr**	AAU, AAC **Asn**; AAA, AAG **Lys**	AGU, AGC **Ser**; AGA, AGG **Arg**	U, C, A, G
G	GUU, GUC, GUA, GUG **Val**	GCU, GCC, GCA, GCG **Ala**	GAU, GAC **Asp**; GAA, GAG **Glu**	GGU, GGC, GGA, GGG **Gly**	U, C, A, G

The abbreviated names of amino acids are as follows: **Ala** = alanine, **Arg** = arginine, **Asn** = asparagine, **Asp** = aspartic acid, **Cys** = cysteine, **Gln** = glutamine, **Glu** = glutamic acid, **Gly** = glycine, **His** = histidine, **Ileu** = isoleucine, **Leu** = leucine, **Lys** = lysine, **Met** = methionine, **Phe** = phenylalanine, **Pro** = proline, **Ser** = serine, **Thr** = threonine, **Trp** = tryptophan, **Tyr** = tyrosine, **Val** = valine.

transfer RNA or **tRNA** for short, is a vital intermediary in this process. Each tRNA can carry a specific amino acid *and* binds to a specific codon—thus, for example, the tRNA that binds to UUU will carry a phenylalanine and no other amino acid. Thus each type of tRNA acts as a kind of 'adaptor', bringing together a codon with its correct amino acid.

By convention, mRNA is translated from left to right as we write its base sequence. The polypeptide chain 'grows', is assembled, one amino acid at a time starting from its N-terminal (amino acid) residue to its C-terminal (amino acid) residue; as it happens, the same direction in which, by convention, we write polypeptide sequences. At the end of the message a special codon, a **stop codon**, occurs—this effectively 'says' end of message. The completed polypeptide chain is released from the ribosome, whereupon it folds up (sometimes in conjunction with other newly synthesized polypeptides) to form a specific protein. This folding involves weak bonds (i.e. non-covalent bonds) and, sometimes, can also involve specific covalent ones that form between two **cysteine** residues (so-called **disulphide bridges**). The route from gene to specific protein product is complete—a route summarized in Figure 5.2. ■

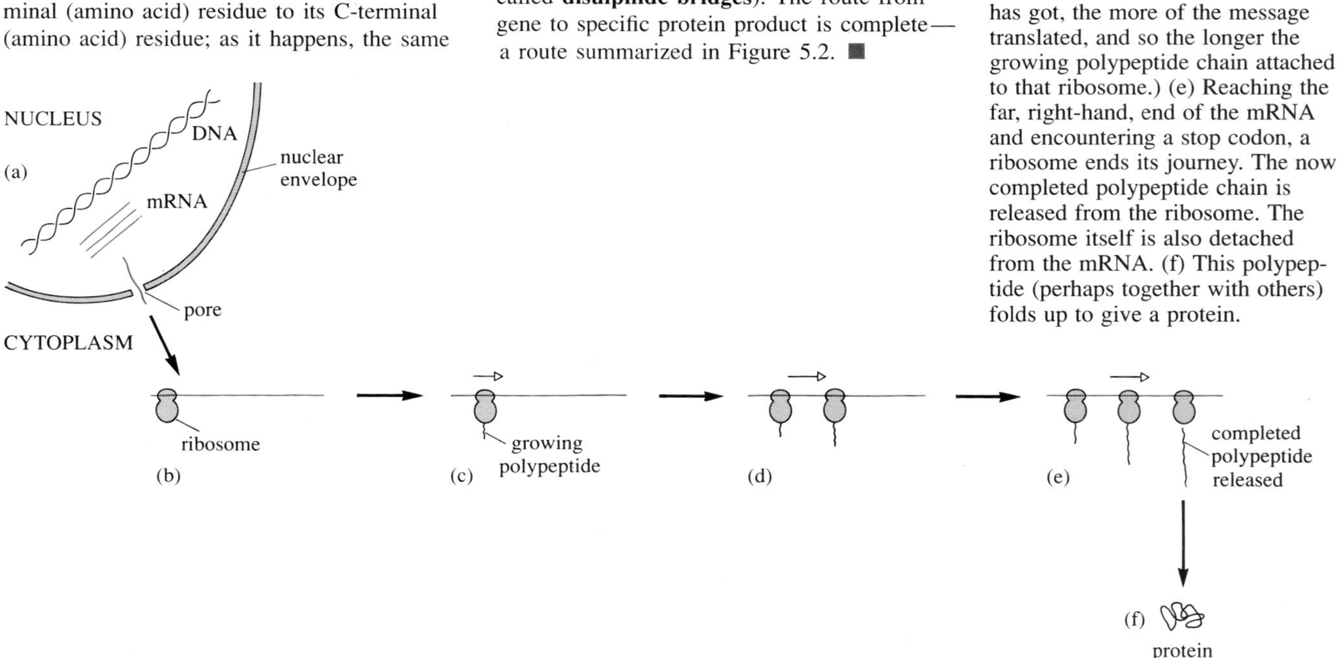

Figure 5.2 The expression of a gene: from DNA to protein. The diagram is for a eukaryote cell; so the DNA is in the nucleus, whereas the ribosomes are in the cytoplasm. (a) For simplicity molecules of mRNA are shown transcribed from one particular gene within the DNA. (b) Molecules of mRNA pass out into the cytoplasm (just one molecule is shown). A ribosome attaches to a molecule of mRNA at its left-hand end. (c) The ribosome travels along the mRNA (see small arrow) translating the message, codon by codon, as it goes. Each codon translated results in another amino acid being added to the growing polypeptide chain attached to the ribosome. (d) Meanwhile, other ribosomes can attach, in succession, to the mRNA molecule at its left-hand end. They too travel along and translate it. Thus each ribosome bears a growing polypeptide chain. (The nearer to the right-hand end of the mRNA a ribosome has got, the more of the message translated, and so the longer the growing polypeptide chain attached to that ribosome.) (e) Reaching the far, right-hand, end of the mRNA and encountering a stop codon, a ribosome ends its journey. The now completed polypeptide chain is released from the ribosome. The ribosome itself is also detached from the mRNA. (f) This polypeptide (perhaps together with others) folds up to give a protein.

To test your understanding of the material in Box 5.1 you should now attempt the following questions.

Question 5.1 The beginning of an mRNA has the following sequence of bases:

AUG UUG UAU UUC UGG GGA UGU GUU UUU UUG...

(The bases are written arranged in triplets for ease of *our* reading.) What would be the sequence of the 10 amino acids in the N-terminal portion of the polypeptide made by translating this stretch of mRNA? (The codons are shown in Table 5.1.)

Question 5.2 Consider the base sequence of the mRNA shown in Question 5.1. Write down the base sequence in the portion of the gene from which that mRNA was transcribed, giving the base sequence for both strands of the DNA.

Question 5.3 Consider again the base sequence of the mRNA shown in Question 5.1. What would be the resulting amino acid sequence in the N-terminal region of the corresponding polypeptide if the 15th base (from the left, as usual) were A instead of G?

Though the specific correspondence between codons and amino acids (i.e. the genetic code) is essentially universal, not all the 'machinery' of gene expression is interchangeable between species. This has important consequences for genetic engineering and so we must consider what expression involves in a little more detail than that shown in Figure 5.2. Let us now look at transcription.

The transcription of a structural gene to produce an mRNA requires the enzyme RNA polymerase, which must attach to the DNA at a *specific* site located just before each structural gene. Such a site is called a **promoter**, as it is the binding of RNA polymerase to it that 'promotes' the subsequent transcription of the adjacent structural gene (Figure 5.3).

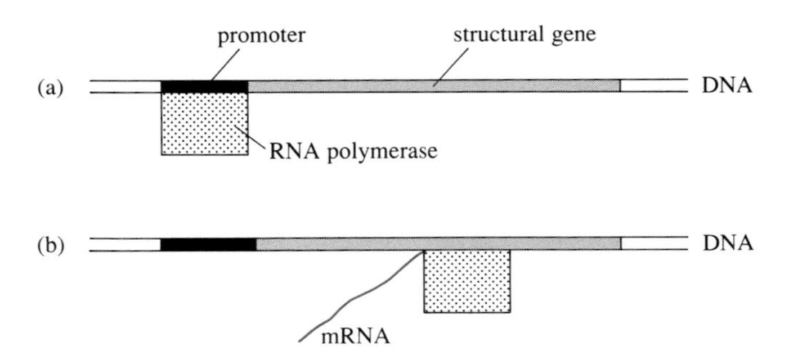

Figure 5.3 The start of transcription. (a) RNA polymerase binds to the promoter. (b) The RNA polymerase works its way along the adjacent structural gene joining nucleotide to nucleotide to form an mRNA as it does so. (See also Figure 5.1 for details of transcription.)

Thus before a structural gene can be transcribed it must have a promoter adjacent to it and that promoter must be 'recognized' by RNA polymerase which can thus bind to it. This ensures that RNA polymerase only binds at correct places on the DNA—that is, only at promoters—and hence only the adjacent bases are transcribed, those comprising adjacent genes. Some promoters are more powerful than others; presumably they are more readily recognized by RNA polymerase and hence promote transcription of the adjacent structural gene more readily.

▷ How might the correct recognition between RNA polymerase and promoters come about? (*Hint*: Think again about what makes DNA specific.)

- Promoters contain specific base sequences. Only where such sequences occur (i.e. in promoters) can RNA polymerase bind.

Though it appears that some promoters can function in more than one species, promoters are not universal; that is, the RNA polymerase of one species may not necessarily recognize promoters of another species. Without this recognition there will be no transcription, no mRNA and no polypeptide product—so even with its own promoter attached, the transferred foreign gene will remain silent, its instruction unexpressed.

So to be expressed, the foreign gene (within a gene-size fragment of DNA) must have its adjacent promoter attached *and* this promoter must be one compatible with the RNA polymerase of the host cell.

It seems that *E. coli* is relatively promiscuous in its recognition of promoters—its RNA polymerase is not particularly choosy as to the species of origin of promoters. For example, *E. coli* can recognize promoters from many other prokaryotes (like itself). Other species, including bacteria such as *Bacillus subtilis*, may be more choosy.

Therefore with *E. coli* as host (as in Figure 4.2), foreign genes from quite a range of organisms might well be transcribed provided that their promoters have also been transferred as part of the gene-size fragments. Having been transcribed, a number of the resulting foreign mRNAs can also be translated in *E. coli* to give their protein products; that is the foreign genes are fully expressed (Box 5.1). However, many eukaryote genes, indeed most, would not be expressed to give protein in *E. coli* and this would be true *even if their own promoters were compatible* with *E. coli* RNA polymerase. Even *if* transcription of the genes occurred in *E. coli*, the mRNA from such genes could not be translated there. To understand one major reason why, we must consider a fundamental difference between the structure of prokaryote genes and that of many eukaryote ones.

5.2 The problem of split genes

5.2

Box 5.2 Split genes

'DNA makes RNA makes protein' is *the* fundamental statement of how the instructions in genes are utilized—so fundamental that in the 1960s it gained the tag of **'the central dogma'**. Subsumed within this was the idea that the instruction in a single gene was *continuous*—a continuous sequence of bases in the DNA. This continuous base sequence was transcribed to give a continuous, and corresponding, base sequence in mRNA. This was translated, in turn, to give a continuous, and corresponding, sequence of amino acids in a polypeptide chain. Experiments bore this out and it seemed a universal fact. Just as the correspondence between particular codons and particular amino acids (the genetic code) was universal so was this—the information in a gene was a continuous base sequence. In 1977 this satisfyingly simple conclusion was given a sharp jolt.

What became apparent was that in many *eukaryotes*, far from being continuous base sequences, many genes were **split**; that is, they were sequences that were interrupted by other sequences *that did not code for protein* (Figure 5.4a). The DNA is not literally split, the polydeoxyribonucleotide chains are continuous, but the *instruction* in the gene is split/interrupted. The non-coding regions within genes were called **introns** as they *intervene* within the coding regions. The coding regions were called **exons**, as they do code for protein and so *exit* from the nucleus (not literally, for it is their mRNA copies that exit) to the *cytoplasm* where they are expressed (Figure 5.2b). (The **cytoplasm** is everything inside the cell except for the nucleus. It can be considered to comprise an aqueous medium, the *cytosol*, with large structures within it, such as mitochondria and ribosomes.) But the mRNA in the cytoplasm *does* contain a continuous sequence of bases all of which are translated to give polypeptide. So there must be an intermediate stage between the transcription of such split genes to give mRNA and the exit of mRNA from nucleus to cytoplasm where it is translated (i.e. a stage between (a) and (b) in Figure 5.2). This stage involves the modification of the newly produced mRNA (the **primary mRNA transcript**, Figure 5.4b) whereby the RNA sequences corresponding to the non-coding *introns* are excised and the sequences corresponding to the coding *exons* spliced together (Figure 5.4c). As this modification of the primary mRNA transcript occurs after transcription, it is sometimes called **post-transcriptional modification**, a process involving an appropriate 'machinery of cutting and splicing enzymes'. (Do not confuse this with the cleaving and splicing used in making recombinant plasmids; Chapter 4.) Thus 'cut and spliced', the mature functional mRNA passes into the cytoplasm where it is translated to give polypeptide (Figure 5.4d). The passage from split gene to protein product is summarized in Figure 5.4.

Not all eukaryote genes are split; that is, some do not contain introns. But many are split and sometimes dramatically so—for example, a gene coding for ovalbumin, a protein in egg white, has seven introns and eight exons; a gene for collagen, a protein in connective tissue, has nearly 40 introns.

Unlike eukaryote genes, with extremely few exceptions, all those of prokaryotes conform to the original, pre-1977, concept of the continuous base sequence—a single uninterrupted DNA base sequence transcribed to give an already mature mRNA capable of immediate translation to give polypeptide. As introns generally do not exist in bacteria, there is no need for post-transcriptional modification of a primary mRNA transcript. The primary transcript *is* the final mature one. Having no need of post-transcriptional modification, prokaryotes do not possess the appropriate machinery for it. ■

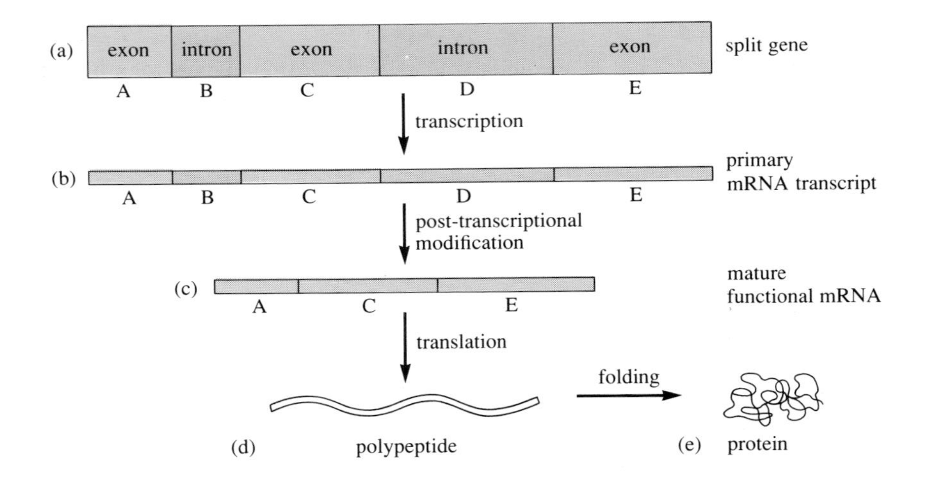

Figure 5.4 From split gene to protein product. (a) A hypothetical eukaryote gene with three exons (in pink) and two introns (in grey) is transcribed to give a primary mRNA transcript (b). This undergoes post-transcriptional modification that removes the portions corresponding to introns and splices together the portions corresponding to exons—steps in the production of (c) a mature functional mRNA. This passes out into the cytoplasm where it is translated to give a polypeptide (d) which folds to give a protein (e). (*Note*: The exons and corresponding mRNA sequences have been given letters merely to clarify the cutting and splicing process.)

To test your understanding of the material in Box 5.2 you should now attempt the following question.

Question 5.4 *Considering only the problem of split genes*, which of the following would you anticipate could be expressed to give their protein product following their transfer into *E. coli*?

(a) a gene from another strain of *E. coli*

(b) a mammalian gene that has no introns

(c) a gene from *Bacillus subtilis*

(d) a mammalian gene with one intron

(e) a wheat gene with four introns.

If a eukaryote gene transferred into a bacterium, such as *E. coli*, happened to be a *split gene* it could not be expressed to give its protein product in the new prokaryotic surroundings. Even *if* transcribed to give a primary mRNA transcript in its novel host, that host will not have the machinery needed to modify that transcript into the mature functional mRNA—the mature mRNA with a continuous base sequence coding for the normal foreign protein product. However otherwise efficient the prokaryote is as a milieu for foreign genes, in order to get this protein product, somehow the genetic engineer is going to have to do this aspect of the job for it.

5.3 Working in reverse — how to by-pass introns

In a nutshell, the problem is this. We have described a method for transferring foreign genes as gene-size pieces of DNA, including split eukaryote ones, into bacteria. Let us assume, for the moment that there is no incompatibility of promoters and that the bacteria are therefore capable of transcribing such eukaryote genes. But, even if the eukaryote genes were transcribed, the bacteria would be incapable of modifying the *primary mRNA transcripts* from those genes so as to remove sequences corresponding to introns, should such occur. As we can transfer *DNA*, the solution is to *transfer DNA (genes) with the introns already removed*. How then can we do this removal?

In effect, the answer lies in allowing the *eukaryote* cells from which the genes come to do most of the task for us. Those eukaryote cells do not remove introns from DNA as such, but they do remove corresponding sequences from the primary mRNA transcripts of those genes. That is, after all, what this post-transcriptional modification is about (Figure 5.4). Thus if we obtain already processed, mature, mRNA from such cells and copy it to produce *complementary DNA*, this DNA will be equivalent to genes without introns, genes that comprise only coding sequences, exons.

Copying mRNA to give complementary DNA is in essence the reverse of what happens normally inside all cells, the process of transcription (Figure 5.1). We can achieve this reverse copying *in vitro* by virtue of an enzyme first discovered in 1970 in cells infected with certain tumour viruses. The enzyme, called appropriately **reverse transcriptase**, has the property that, when presented with an mRNA and a source of (deoxyribo)nucleotides, it will assemble a DNA strand from these nucleotides with a base sequence complementary to the mRNA (Figure 5.5a).

▷ What will ensure the complementarity of base sequence between the mRNA and the single-stranded DNA produced?

▶ The mRNA acts as a *template*; each base in the mRNA will ensure that its complementary base is inserted in the assembling DNA strand. The process is the reverse of transcription (Figure 5.1) but the rules of base complementarity hold—A, G, C and U in mRNA with T, C, G and A in DNA, respectively.

This *single-stranded* DNA molecule can then be used as a template to produce a DNA strand complementary to *it* (this uses another enzyme, **DNA polymerase**). The net result is a double-stranded DNA molecule the base sequence of which corresponds to the mature mRNA (Figure 5.5b), a so-called **complementary DNA** (usually called

cDNA). We have produced an 'artificial gene' that contains only a continuous coding sequence (only exons)—in effect, what the gene corresponding to the mature mRNA would be if split genes did not exist.

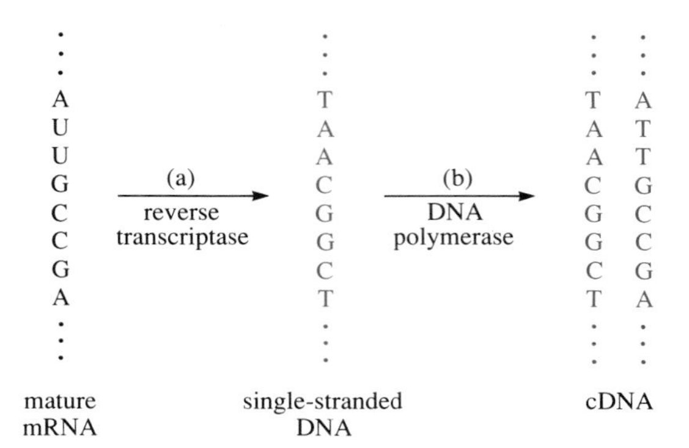

Figure 5.5 Producing cDNA. (a) Using mature mRNA, shown in black, as a template, reverse transcriptase yields a single-stranded complementary DNA, shown in red. (b) With DNA polymerase, this single-stranded DNA acts as a template to produce *its* complementary strand (also shown in red). The net result is a double-stranded DNA with base sequence corresponding to that of the mature mRNA. (The sequences shown are hypothetical and partial; the dots indicate many further bases that are not shown.)

Utilizing reverse transcriptase thus provides an alternative strategy to cloning eukaryote genes isolated as fragments of foreign DNA (using restriction enzymes as in Figure 4.2d). Instead foreign cells are taken and the mRNA extracted from the cytoplasm (Box 5.2).

▷ Why cytoplasmic mRNA?

▶ We need *mature* (functional) mRNA, that is mRNA from which all intron-corresponding sequences have been excised. Such mRNA is in the cytoplasm (where it can be translated); nuclear mRNA would include primary transcripts containing sequences corresponding to introns (Figure 5.4).

This cytoplasmic mRNA is then used with reverse transcriptase and DNA polymerase to produce cDNA molecules (Figure 5.5). The cytoplasmic mRNA will in fact be a mixture of different mRNA molecules. Each type of mRNA will correspond to a different gene (at least to the exons of a split gene) and code for a different polypeptide product. Thus, after reverse transcriptase and DNA polymerase, we will end up with a *mixture* of different cDNA molecules. Each of these different cDNAs corresponds to the entire *coding* region of a different gene, that is, equivalent to a gene without its introns. Thus each cDNA is automatically 'gene-size' and hence suitable for splicing into a plasmid.

Once spliced into plasmids, cDNAs can be taken up into bacteria—one cDNA per plasmid and one plasmid per bacterium. Each bacterium can then be allowed to grow and divide to give a clone. We thus end up with a series of bacterial clones, each clone containing a different cDNA, including, hopefully, a clone with the cDNA we desire (i.e. with the exons of the desired gene). The overall production and cloning of cDNAs is shown in Figure 5.6a–f).

If you compare it with Figure 4.2, you will see that from cDNA on the procedure is essentially the same to that developed for cloning restriction enzyme fragments of foreign DNA (Figure 5.6c–f, matching Figure 4.2e–h). And, in Figure 5.6 we have included an additional step to show the expression of the cloned cDNAs to give protein. This step (Figure 5.6g) is analogous to that in Figure 4.2i.

But, as described so far, would this expression of cDNAs to give protein actually occur? Consider what expression of a cDNA in *E. coli* must involve—first transcription of the cDNA to give mRNA, then translation of this to give polypeptide (Box 5.1).

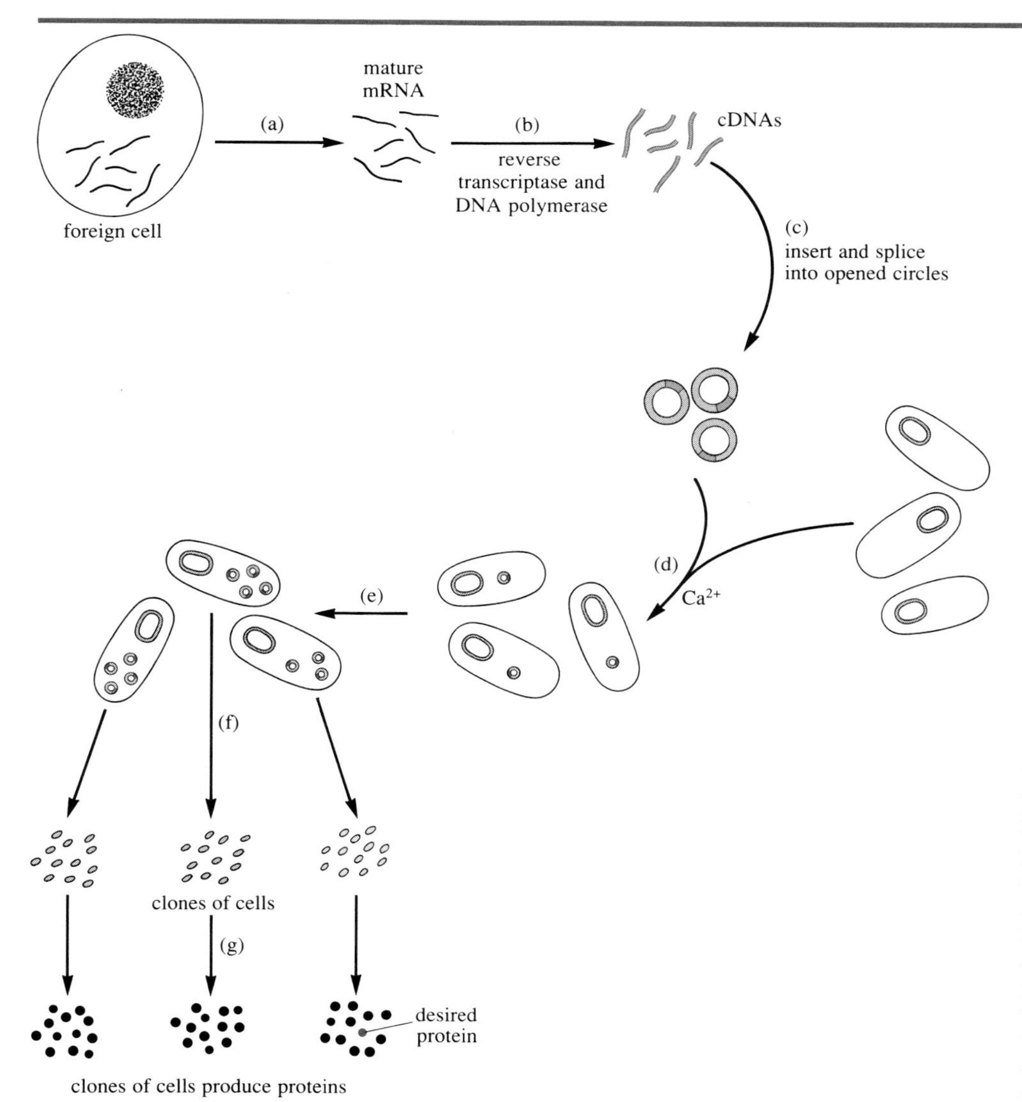

Figure 5.6 Cloning cDNAs. (a) a mixture of mRNAs is extracted from the cytoplasm of foreign cells. (b) Using reverse transcriptase and DNA polymerase, the mRNA is copied to give a mixture of cDNAs (in red). (c) Each cDNA is spliced into a previously opened plasmid circle (in grey) to produce a recombinant plasmid. (d) In the presence of calcium ions, individual recombinant plasmids are taken up by individual plasmid-free *E. coli* —one plasmid per cell. (e) The recombinant plasmids replicate. (f) *Each* cell containing recombinant plasmids is allowed to grow and divide to form a separate specific clone; among all these different clones is the one whose cells contain the desired cDNA. (g) The clone whose cells contain the desired cDNA produces the protein product of that cDNA.

But think back to what we have done. We have taken the cDNA route (rather than the gene-size fragment one; Figure 4.2) to by-pass the problem of introns. Thus in transferring cDNAs into *E. coli* we have transferred *just* the coding sequence of foreign eukaryote genes. Unlike the 'native' gene plus promoter (i.e. as transferred on a gene-size fragment), the cDNA will have no promoter attached. So for a cDNA to be transcribed to give mRNA in *E. coli* we must 'provide it' with a promoter. What is more, this promoter must be compatible with *E. coli* (i.e. with *E. coli* RNA polymerase; Section 5.1). The obvious solution to both problems is to provide the cDNA with an *E. coli* promoter. This can be done by using what is known as an *expression vector.*

5.3.1 Using expression vectors with cDNAs

As the name suggests, an **expression vector** is a vector that permits the expression of foreign genes in their novel host surroundings. An expression vector can be a plasmid that has been manipulated *previously* so as to contain a structural gene (plus a little adjacent DNA) that is expressed normally in the host cells (say, *E. coli*). It must therefore have also its adjacent promoter that is competent in that host (Figure 5.7a); i.e. compatible with the host's RNA polymerase (an *E. coli* promoter is the obvious candidate).

▷ How might such 'manipulation' of a plasmid to create an expression vector have been achieved?

▶ By the standard techniques of gene cutting and splicing, much as outlined in Figure 4.2.

The expression vector can be opened and the foreign cDNA spliced in (by techniques analogous to step (c) in Figure 5.6) *within* the structural gene (Figure 5.7b and c). This plasmid is then taken up by *E. coli* (much as in Figure 5.6d) and clones produced. Having its own promoter, transcription of the *expression vector* structural gene can commence. But transcription *continues into and right through the foreign cDNA* until it reaches a signal (say, in the vector DNA) that in effect says 'terminate transcription' (Figure 5.7d and e). This yields a *hybrid* mRNA (Figure 5.7e). This is translated to give a *hybrid* polypeptide (Figure 5.7f) wherein the first part (the N-terminal part; Box 5.1) corresponds to the first part of the *expression vector* structural gene and the rest corresponds to the whole of the foreign cDNA. (Translation will stop at the stop codon corresponding to the end of the foreign cDNA.) Hybrid polypeptide can be isolated from the cells and treated (e.g. chemically; Figure 5.7g) to remove the N-terminal section leaving the desired portion of polypeptide—just that corresponding to the foreign cDNA.

Thus if we required our cDNA cloning technique to give us not just transfer but also expression (Figure 5.6g) of the cloned genes (i.e. cDNAs), we should use a plasmid that is an expression vector, in the stage shown in Figure 5.6c.

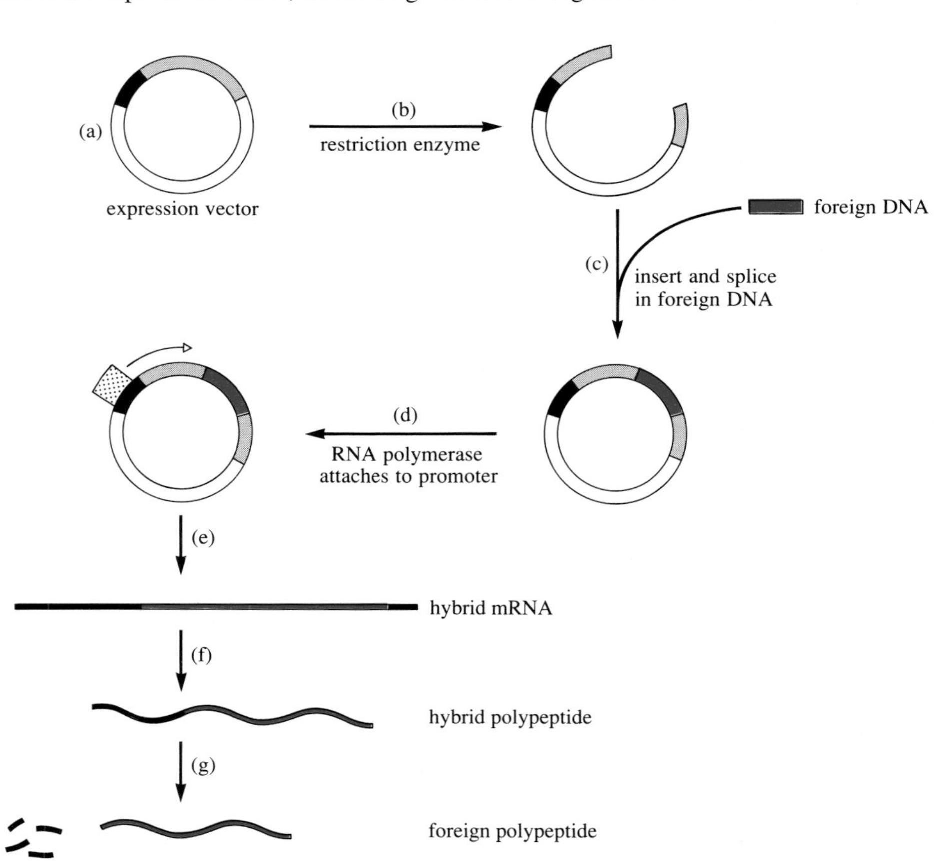

Figure 5.7 Using an expression vector for expressing cDNAs. (a) The structural gene (in grey), which is part of the expression vector, has its own adjacent promoter (in black). (b) The expression vector is treated with a restriction enzyme that cleaves it at a single site, one *within* the structural gene. (c) The foreign cDNA (in red) is inserted into the opened expression vector (within the structural gene) and the join sealed with ligase. This gives a re-circularized plasmid (expression vector) containing the foreign cDNA (hence it is a recombinant plasmid). (d) When this plasmid is taken up into *E. coli*, RNA polymerase attaches to the promoter of the expression vector structural gene and transcription commences. (e) It continues right through the spliced-in foreign cDNA. This gives a hybrid mRNA (the part corresponding to expression vector structural cDNA sequence is in black, that to foreign cDNA sequence is in red). (f) The hybrid mRNA is translated giving a hybrid polypeptide (similarly colour-coded). (g) The experimenter isolates the hybrid polypeptide and, using, say, a suitable chemical, cleaves off the N-terminal section, leaving the desired polypeptide product of the foreign cDNA intact.

5.3.2 Using the right source for the mRNA

There is one potential snag with using cDNA cloning. This relates to where we get the cDNAs from, or, more correctly, where we get the mRNAs that we use to produce cDNAs from. Remember that we wish to clone a *specific* foreign gene (here represented by the corresponding cDNA) and get its protein product expressed in bacteria. So success will depend on starting with a mixture of mature mRNA that contains at least some mRNA molecules corresponding to the *specific* foreign gene (in fact, to its exons) that we are interested in. But the only mRNAs present in the mixture extracted from the foreign cells will be ones coding for proteins *being made in that cell, at that time* (because, as you will recall, not all the genes in a cell are expressed at any one time; Box 5.1). Therefore we must be sure that the foreign cell chosen is expressing the gene we are interested in and therefore will contain some of the appropriate mRNA among the total mRNA mixture. Thus, for example, if we wish to obtain a cDNA that will code for hen ovalbumin, an egg-white protein, it makes sense to extract mRNA from chicken oviduct cells rather than, say, from red blood cells. Choosing an appropriate cell can have another advantage.

If the protein happens to be a major component of the total protein being synthesized in that cell, then it is likely that there is plenty of mRNA for that protein present in the cell cytoplasm. This means that when we extract the mixture of mRNAs from that cell a reasonably large proportion of the mixture will be the mRNA that we want. For our case in point, hen ovalbumin, there would be about 100 000 molecules of the ovalbumin-coding mRNA per oviduct cell, as compared to rarer proteins where there may be just tens or fewer molecules of each of their mRNAs per same cell. When put through the reverse transcriptase and DNA polymerase treatment, such a mixture of mRNAs would result in a population of cDNA molecules in which the cDNA corresponding to the common mRNA (say, that for hen ovalbumin) would be similarly well represented.

When we subsequently inserted these cDNAs into plasmids and thence into bacteria (Figure 5.6c and d), there would be a relatively high proportion of bacteria that contain the required cDNA; hence a high proportion of the clones grown from them would contain the desired cDNA (Figure 5.6f). Where the protein and therefore the mRNA are not in abundance in a cell, the same cDNA technique will still work. However, the number of clones containing the desired cDNA (and, where expressed, its protein product) may be very many fewer.

We have now finished our description of methods for transferring foreign genes (or the cDNA equivalents) into *E. coli* and getting those genes expressed, whether they are non-split genes (prokaryote and some eukaryote; summarized in Figure 4.2) or split eukaryote ones (summarized in Figure 5.6). Our path from foreign gene to foreign protein in bacteria is complete. Or is it?

Summary of Chapter 5

1 To obtain the polypeptide (protein) product of a foreign gene in bacteria that gene must be *expressed*.

2 Gene expression involves two stages: *transcription* to produce a *messenger RNA (mRNA)*; and *translation*, the reading (decoding) of the mRNA to produce *polypeptide*. Polypeptide then folds up to yield the ultimate protein product.

3 Transcription depends on having an appropriate *promoter*, one compatible with the *RNA polymerase* in the host cell. The promoter is the region in the DNA *adjacent to a structural gene* where *RNA polymerase binds* prior to transcribing the structural gene.

4 Most eukaryote genes are *split genes*. Split genes contain coding regions of DNA, *exons*, interspersed with non-coding regions, *introns*. These genes *cannot be expressed* in bacteria.

5 In eukaryotes, mechanisms exist that 'cut and splice' the *primary mRNA transcripts* from split genes. This *post-transcriptional modification* gives *mature functional mRNA* that passes into the cytoplasm where it is translated to give polypeptide. Split genes do not occur in prokaryotes (except extremely rarely) and so neither does the machinery for such post-transcriptional modification.

6 The intron problem can be by-passed by transferring *cDNAs* instead of gene-size pieces of foreign DNA.

7 A cDNA is a *complementary DNA* made by using a *mature cytoplasmic mRNA* as a *template*. The first step, utilizing the enzyme *reverse transcriptase*, produces a single-stranded DNA complementary in sequence to the mRNA template. This single-stranded DNA is converted to a *double-stranded cDNA* using another enzyme, *DNA polymerase*.

8 A *mixture of mRNAs* extracted from the cytoplasm of a suitable foreign cell can be used to produce a *mixture of cDNAs*. This mixture is then used as the 'source of genes', individual cDNAs being spliced into plasmids to create recombinant plasmids. The recombinant plasmids are then taken up into *E. coli* cells which are then grown to produce individual clones.

9 To get expression of cloned cDNAs we must provide a compatible promoter adjacent to the cDNA. This can be done using an *expression vector*. An expression vector can be a plasmid that contains a promoter compatible with the RNA polymerase of the host species.

Question 5.5 Figure 5.8 shows a simplified diagram of the path from gene to protein product in a prokaryote cell.

(a) Match letters A–E in Figure 5.8 to the following items: mRNA; polypeptide; growing polypeptide chain; RNA polymerase; ribosome.

(b) Which of the items in (a) would be present in a eukaryote cell?

(c) Which of the following items *must always* also be operating or present in this pathway if the cell is a eukaryote one: a nuclear envelope; the machinery of post-transcriptional modification?

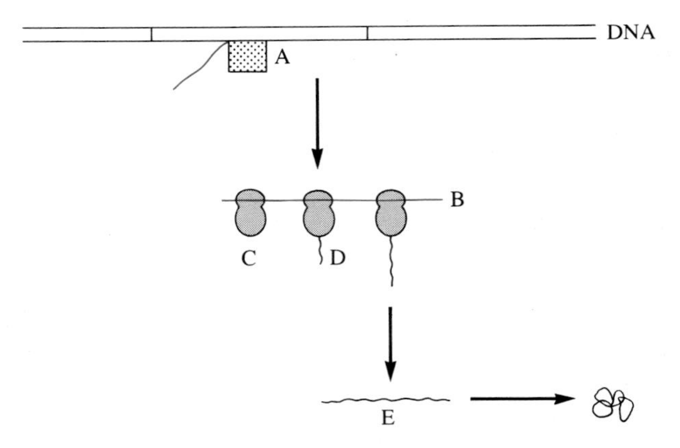

Figure 5.8 For use with Question 5.5.

Question 5.6 Which of the following statements are true and which are false?

(a) If we wish to transfer eukaryote genes into bacteria then we must use cDNA cloning, not cloning of gene-size pieces of DNA produced by restriction enzymes.

(b) If we wish to get the production of eukaryote proteins in bacteria then cDNA cloning is the appropriate technique, not cloning of gene-size pieces of DNA produced by restriction enzymes.

6 Getting the right clone

We have finished our description of how to transfer foreign genes or their cDNA equivalents into bacteria and get their protein products produced there. We now need to consider how to get *specific* foreign protein products produced in bacteria; something we do in this chapter. Once again you will need Figure 4.2 and preparing your own chapter summary should help you to understand and memorize the main principles. This chapter is a relatively long one and quite involved. You should read all of it but, if short of time, you can regard Section 6.4.3 on how to obtain gene probes as 'optional'.

Right at the outset of considering methods for transferring genes into bacteria (Section 4.2) we raised a question: how in the process does one actually transfer the *desired* gene(s)?

Protocols along the lines of those in Figures 4.2 and 5.6 achieve the cloning of gene-size pieces of DNA or cDNAs, respectively—the cells of each clone containing a particular gene (within a piece of foreign DNA) or its cDNA equivalent. Using the appropriate protocol we may also get such genes or cDNAs expressed. But we are after the transfer of one particular gene. How do we know where our desired gene (or cDNA) is, *in which clone*?

We do not in fact separate the foreign DNA fragments or cDNAs *before* cloning (i.e. before making recombinant plasmids—step (e) in Figure 4.2 or step (c) in Figure 5.6); in other words, pick out our desired gene/cDNA, and then clone it alone. Given that there might be a huge number of different gene-size fragments or cDNAs it would be virtually impossible to do so—physically, one DNA fragment is very much like another of a similar size and it would be very difficult to separate them cleanly. And then there would still be the problem of identifying the desired DNA fragment or cDNA.

In essence, what we actually do is to *clone all the fragments of foreign DNA* (Figure 4.2e and beyond) or *all the cDNAs* (Figure 5.6c and beyond). Each DNA fragment/cDNA will then end up in a separate bacterium (Figure 4.2f or Figure 5.6d) and hence, ultimately, in a separate clone (Figure 4.2h or Figure 5.6f).

Using protocols along the lines of that in Figure 4.2, we can clone the *entire* genetic material, the **genome**, of a foreign cell in bacteria. Each individual gene from the foreign cell, or a small cluster of neighbouring genes present in a single gene-size fragment, will end up present in a separate bacterial clone. This gives us what is sometimes termed a **gene library**; all the genes of the foreign cell are distributed among the bacterial clones and each bacterial clone contains a particular gene (or small cluster of genes) from the foreign organism.

Analogously, the cDNA route (Figure 5.6) allows us, *in effect*, to clone all the genes (minus introns) corresponding to mRNAs being made in the foreign cell at the time of mRNA extraction. This gives us a **cDNA library**.

But, as has been pointed out, in each case we have 'a library without a proper index'. Because we do not know where particular 'books' are located, which gene/cDNA is in which clone. As we are after one particular gene/cDNA we must then examine the library—**screen** the individual clones to find the one that contains the desired gene/cDNA.

Thus, in summary, these protocols do *not* involve precise pre-selection and targeting of the desired gene/cDNA to be transferred. In effect, we transfer many genes, one or a few at a time (as DNA fragments or cDNAs), produce separate clones of bacteria each containing a different foreign gene(s)/cDNA, and *afterwards* screen, search, for the desired clone, that containing the desired gene/cDNA. Such protocols are called appropriately **shotgun techniques**. We take a relatively blind shot (using gene-size pieces of DNA or a mixture of cDNAs) at the target (the *desired* gene; not to be confused with the 'target sequences' of restriction enzymes) and look for hits afterwards (screening the clones).

6.1 How many clones must we screen?

The blinder the shot, the harder you have to look to find whether you have hit the target and to find where the target is.

To understand this, say we were using a cloning protocol like that in Figure 4.2 to obtain one particular gene from among the many in a foreign organism — we simply chop the foreign DNA into gene-size fragments, clone these in bacteria and screen the clones for the particular fragment that bears the desired gene. But the numbers of clones to be screened could be huge.

For example, consider the human (haploid) genome. This comprises in total some 2.9×10^9 base pairs. Now, say, each of the *different* gene-size fragments produced following cleaving a preparation of human DNA with a particular restriction enzyme (Figure 4.2d) averages around 5×10^3 base pairs. If we clone these fragments in bacteria, we should expect around $2.9 \times 10^9/5 \times 10^3$, that is 5.8×10^5, over half-a-million, *different* clones and we would need to screen each one to find the one we require. But in fact we would need to screen more clones than this. To understand why, consider what we would actually do to foreign DNA in such a cloning experiment.

In practice, our sample of DNA would come not from one cell but be extracted from very many identical cells. So the sample in the test-tube would contain *very many identical copies* of the same type of DNA. This particular DNA will have within it a number of target sites for the restriction enzyme that has been chosen. On treatment with the enzyme, *every* one of the very many identical copies of DNA will be cleaved at these same sites. This results in a number of *different* gene-size fragments of DNA, *each* copy of the DNA yielding the same mix of fragments. The net result, the *restriction enzyme digest*, is a number of *different* gene-size fragments, each different (type of) fragment present in the test-tube very many times. Figure 6.1 should help clarify this.

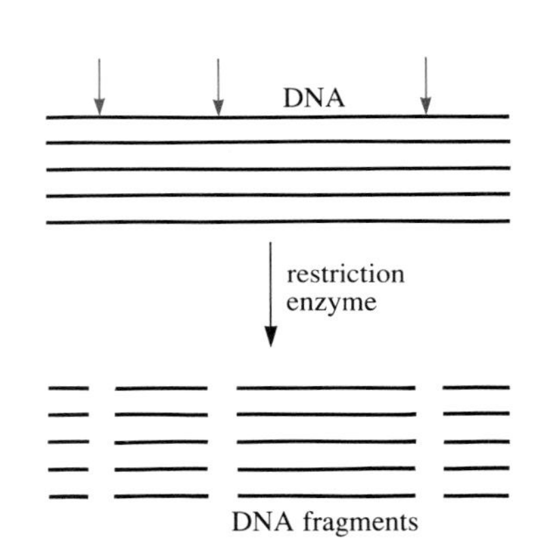

Figure 6.1 Fragments generated by treating a hypothetical sample of DNA with restriction enzyme. We consider just one type of DNA molecule present very many times (here, however, just five, for simplicity). The target sites for the restriction enzyme to be used are indicated by the red arrows. On treatment with the enzyme each of the molecules of DNA yields the same mix of fragments; each type of fragment is thus present very many times (here, of course, just five).

It is this mix of fragments — many types, each present many times — that we then clone (Figure 4.2e and beyond). Then we should end up with many clones *for each different type of fragment*; and, in theory, our desired clone should crop up once in every half-million clones. But screening *just* a half-million clones to find one we want would be pushing our luck. (To follow this, consider trying to toss a coin to get a tail. In theory this occurs 1 in 2 tosses. But you might need several more in practice to 'guarantee' success.) There are statistical ways of calculating how many clones we should need to screen. In our example of human DNA, screening around 1.3×10^6 clones would give us a 90% chance of finding one with any particular desired gene.

Using cDNA cloning, the 'shot' may not be so blind — a modern shotgun rather than an old-fashioned blunderbuss perhaps. Firstly, there may well be many fewer different cDNAs than there would be different DNA fragments from the restriction enzyme protocol.

▷ Why?

▶ In the restriction enzyme protocol (Figure 4.2), we are chopping the *entire* genome of the organism into gene-size fragments. In cDNA cloning we are generating gene-size cDNAs by copying mRNA. In any particular type of cell not all the genome is being transcribed to mRNA (Box 5.1). Therefore the total number of different mRNAs, and thereby cDNAs, generated from one cell type represents a subset, a fraction, of the total genome of the organism.

Secondly, as mentioned already in Section 5.3.2, *in some circumstances* the particular mRNA we are after may be an abundant one in the type of cell from which we isolate it. Therefore, the corresponding cDNA will also be relatively abundant. When the cDNAs are then cloned in bacteria such an abundant cDNA will by chance turn up in relatively many bacteria and hence relatively many clones (Figure 5.6c–f). The number of clones we need screen to find it will be thus proportionately fewer.

So much for the problem of how many clones we must screen. But how is the actual screening done? How can we examine what might be a very large number of individual clones of bacteria to find the one(s) with the desired gene/cDNA? To answer this, we must digress slightly to consider how to screen bacteria in general.

6.2 The microbial haystack — how to find a needle

You may recall that when we were considering the breeding of microbes, such as bacteria, in Section 3.3, we argued that the problem was not so much one of obtaining useful variants (mutants) but more one of identifying and isolating them — finding and selecting the desired variety from among the vast 'microbial haystack', *screening* the haystack.

The first trick is that we do not literally examine or test individual bacteria. We take the population to be screened and allow each bacterium in it to grow and divide to produce its own clone of descendants. We then can screen each clone — it is easier to test a whole population of genetically identical bacteria (a clone) for some shared characteristic (say, the ability to produce a certain chemical) than to test the single minute ancestor. But how can we get each bacterium to produce a clone that is *physically apart* from other clones, to allow us to test each clone as a separate entity?

One convenient way to do this is to take a portion of the population to be screened, say a portion that contains a thousand or so individual cells, and spread these cells out onto the surface of sterilized solid jelly that contains the nutrients needed for bacterial growth (Figure 6.2a). This jelly is contained in a special covered dish called a Petri dish. In this technique, usually called **plating**, the spreading separates each invisible cell from all others. Left for a while, typically overnight (remember the rapid rate of reproduction of bacteria; Section 3.2), each bacterial cell will grow and divide where it 'sits' on the jelly and thereby produce its clone as a separate visible mound, called a **colony** (Figure 6.2b). Each clone, that is each colony, can then be tested for the characteristic being looked for. To screen a whole population of bacteria many Petri dishes may be needed, as there is a limit to how many colonies can be grown well-separated on one dish.

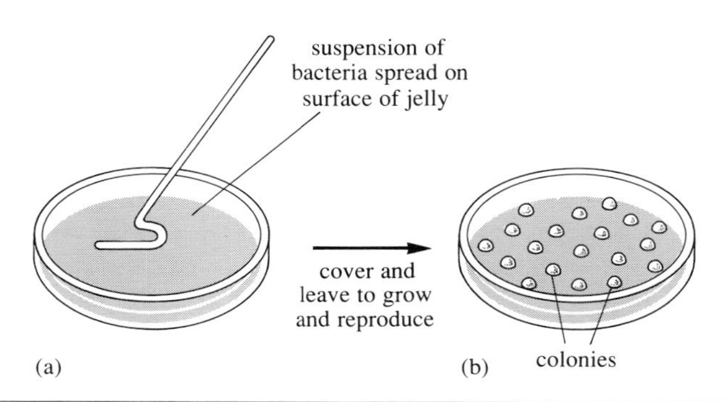

Figure 6.2 Plating. (a) A few drops of suspension are spread in a thin layer over the surface of the sterilized nutrient jelly in a sterilized Petri dish. The dish is then covered with its lid, to keep out any other microbes. The liquid layer is soon absorbed into the jelly. Trapped on the surface each cell grows and divides where it 'sits'. (b) By the time its descendants have reached around a million in number, they are visible as a single colony. Each colony must be descended from a single cell and thus constitutes a population of genetically identical individuals (a clone). (For simplicity, only a few colonies are shown.)

6.3 Screening the gene product

Assume that we have transferred all the gene-size pieces of foreign DNA or cDNAs, whichever is appropriate, via recombinant plasmids into bacteria (Figure 4.2f; Figure 5.6d). The bacteria have been plated onto nutrient jelly (Figure 6.2a). Left to grow (Figure 4.2g–h; Figure 5.6e–f), the bacteria will yield separate clones, each clone visible as a separate colony (Figure 6.2b).

For simplicity's sake, *assume that all the bacteria growing on the jelly are descendants of ones that received a recombinant plasmid*. Actually, not all bacteria from protocols like those dealt with would receive plasmids, on average only about 1% of *E. coli* cells would take up plasmids in such an experiment (Figure 4.2f; Figure 5.6d) and not all such plasmids need be recombinant ones (some without foreign DNA would slip through). Nevertheless, there are tricks, which we need not consider, to ensure that the only colonies that actually grow are from cells that contain plasmids and, further, to tell which of those colonies contain cells with *recombinant* plasmids. It now remains to screen each colony to find those in which the cells contain, not just any recombinant plasmid, but one with the desired foreign gene/cDNA.

Let us assume further that the foreign genes/cDNAs will be expressed to produce their protein products in the bacterial cells. (That is, either the foreign DNA fragment has its own, compatible, promoter attached *or* the cDNA has been inserted into an expression vector; Sections 5.1 and 5.3.1.) It is therefore not necessary for us to screen the colonies for the *desired* gene/cDNA itself, something that may sometimes be relatively tricky to do (Section 6.4). Instead, we can screen for the presence of its protein product, something that can often be very much easier. (If the protein product is present in a colony, then obviously the gene/cDNA for that product must be present too!) Of course, this presupposes that we have a suitable test for the presence of the protein product in question.

For example, let us suppose that the protein product of the gene/cDNA in question is an enzyme.

▷ How in general could we test a colony for the presence of any particular enzyme?

▶ Enzymes are **catalysts**. That is they speed up reactions. By definition those reactions must be able to occur in the absence of the catalyst (enzyme). But the catalytic effect of enzymes is so great that, to all intents and purposes, the reactions that they catalyse would be so slow in the absence of the enzymes as to be virtually undetectable. You will recall that a key feature of any enzyme is its **specificity**. That is each enzyme has a single **substrate**, or a small range of chemically closely related substrates, with which it can interact, catalysing the particular reaction yielding the (also specific) reaction products. Thus, by providing each colony with the specific substrate of *just the enzyme in question* and looking for the products of the enzymic reaction, we can see which of the colonies is capable of carrying out the reaction at a rapid rate and hence must contain the enzyme. (All the cells in a single colony are genetically similar, i.e. are members of the same clone. So if the enzyme is present, all of the cells in the colony must contain it and hence contain the respective gene/cDNA.)

How precisely we provide the substrate and look for the products will depend on the enzyme in question. In some cases it may be very simple. For example, if a substrate is available that undergoes a colour change on reaction we may be able to 'flood' the Petri dish containing the recombinant colonies (a dish like that shown in Figure 6.2b) and look for those colonies that undergo the colour change.

Of course, the protein product of the foreign gene/cDNA may not be an enzyme or may be an enzyme for which no convenient assay is available. In such circumstances it may be possible to use an **antibody** to detect the protein product.

Put simply, antibodies are a class of proteins produced by the immune systems of vertebrates. That is, the immune system can detect certain substances as being 'non-self' and in response produces specific proteins, called antibodies, that can bind to these 'foreign' substances (which themselves can be proteins). This is just part of the protective system that acts to neutralize the effects of foreign substances (such as might be components of invading bacteria or viruses, for example). Each antibody is specific for a particular foreign substance. We can exploit this property of the immune system to produce antibodies *specific against particular proteins*. We can then isolate and purify these antibodies and use them to detect such proteins. How precisely such antibodies are produced and used does not matter for our present discussion. (Further discussion of antibodies can be found in Section 9.2.1.)

What is important is that we can produce an antibody *specific* for the protein product of the desired foreign gene/cDNA; that is, it specifically binds to that protein and that one alone. We then use this antibody to screen the colonies for the presence of the protein and, once again by implication, the presence of the desired foreign gene/cDNA.

Thus, whether an enzyme or not, there are methods for detecting the specific protein product of a foreign gene/cDNA; thereby locating that product, and hence that gene, to a particular colony, a particular clone, of cells. Our journey from specific foreign gene (whether via gene-size fragments or cDNA) to its expression and location in a particular clone of bacteria is now indeed complete.

Activity 6.1 *You should spend up to 30 minutes on this activity.*

This activity draws on material discussed in Chapters 4, 5 and 6.

α-interferon is a protein that occurs in vertebrates that has the property of inhibiting the growth of viruses. It is not an enzyme. α-interferon is produced by white blood cells called leukocytes.

Imagine that to obtain large amounts of human α-interferon, you wish to transfer interferon-producing capacity to *E. coli*. Assume that you know nothing of the detailed structure of human α-interferon nor of its gene.

In outline only, give a possible protocol for such a project. Design the route so as to minimize the number of bacterial clones to be screened; suggest *two* possible tests for screening the clones for interferon production. (*Hint*: although α-interferon is not an enzyme, it does have defined biological activity.)

6.4 Screening for genes directly

In many of the instances of genetic engineering with which we are concerned in this book we need a transferred gene to be expressed to give its protein product (insulin, thaumatin and interferon, for example). We have thus detailed some of the requirements for such expression to occur. We have also shown at the same time how expression can be the basis of locating (screening for) a desired foreign gene. Naturally, for such screening techniques to work, *the foreign gene must be expressed* in its new, *E. coli*, surroundings. Expression may be achieved if the gene is not split and has been transferred as a DNA fragment (Figure 4.2) containing a compatible adjacent promoter (Section 5.1). Or it may be achieved by using the cDNA route (Figure 5.6) with the cDNA inserted into a suitable expression vector (Section 5.3.1). But what if we have no suitable technique to detect the protein product of the desired gene? And, what if the foreign gene is not expressed? Indeed, say we are not even interested in its expression. Let us take some hypothetical cases.

If we wish to get expression of split genes in *E. coli*, then a cDNA route is the one to use. But say we are not interested in expression but wish in fact to study the detailed structure of such a eukaryote gene, *including its introns* (Box 5.2). Then a cDNA route is no use, as a 'cDNA equivalent' of a gene is that gene *minus* its introns (Section 5.3). Yet studying the structure of a particular eukaryote gene is greatly facilitated by cloning just that gene in *E. coli*. So it is still useful to clone it.

▷ Which type of cloning route could we use?

▶ To get a eukaryote gene *with* its introns, a restriction enzyme route could be used (Figure 4.2).

But using this route, even if the gene is transcribed in *E. coli* (say its promoter is compatible), the primary mRNA produced could not be modified (Box 5.2) and hence translation could not occur—no expression. So without expression how do we detect the desired transferred gene? How do we screen the clones for the foreign gene when we *know* it cannot be expressed *in principle*?

Other cases of non-detection may occur where we do *not* know why. For example, a gene-size fragment of DNA may or may not have its own promoter attached. We rarely know in advance. So if expression does not occur, how do we know whether the gene has been transferred and not expressed *or* just not transferred at all (perhaps none of the clones contain that desired gene)? Likewise with a cDNA cloned via an expression vector (Figure 5.7); if no expression is detected, does this mean that the expression vector has failed *or* that no transfer of the desired cDNA has occurred?

It would be useful to have techniques to detect the transfer of a desired gene *irrespective* of whether that transfer permits expression in the new surroundings or not. Such techniques do exist and depend on *detecting the gene itself,* rather than on detecting its protein product (*if* any such occurs).

6.4.1 *Screening DNA* en masse

As will become evident, screening techniques that detect genes directly actually involve 'testing' the DNA in the clones (colonies; Figure 6.2) resulting from gene transfer (of either gene-size fragments, Figure 4.2; *or* of cDNAs, Figure 5.6). This means that DNA must be extracted from each colony and this in turn would mean destroying the cells. But, ultimately, having found the colony with the desired gene, we want those cells, not their debris! The obvious answer is that we must take a sample of each colony (just some of its cells), extract DNA from the sample and test it; thus leaving some of each colony intact. But sampling may itself seem problematic—remember that shotgun techniques may yield very large numbers of clones (Section 6.1) and hence many Petri dishes, each with many colonies. To sample each colony *separately* could be a very forbidding prospect. Fortunately, a neat trick allows us to sample and test the DNA from each colony *en masse*, a Petri dish at a time. This trick involves *replica plating*.

As the term suggests, **replica plating** means making an exact copy, or *replica*, of the colonies. To do this for DNA testing, we use a porous disc, a filter, made of nitrocellulose. The colonies to be screened are grown on nutrient jelly in a Petri dish (the **master plate**, Figure 6.3a). The nitrocellulose disc is then brought into 'face-to-face' contact with the surface of the master plate — that is, the two surfaces, colony-bearing one and nitrocellulose disc, are lightly pressed together (Figure 6.3b). This transfers a small portion of each colony onto the nitrocellulose disc — and *the array of the portions transferred on to nitrocellulose matches that of the colonies on the original dish.* The disc is then removed (Figure 6.3c) giving a replica, on nitrocellulose, of the master plate. (The transferred portions of clones, on this replica, can then themselves be grown by placing the disc on fresh nutrient jelly.) All destructive procedures (e.g. extracting DNA) can then be done using the *replica on the nitrocellulose disc* (Figure 6.3d), retaining the master plate intact.

The replica is then treated with sodium hydroxide, a strong alkali. This has two effects. First, it breaks open the cells, thus 'exposing' all their DNA (chromosomal and recombinant plasmid). Secondly, it causes the individual strands in each double-helical molecule of DNA (Box 4.2) to separate from each other, resulting in *single-stranded* DNA. This is critical to our screening procedure, as will be seen below. The

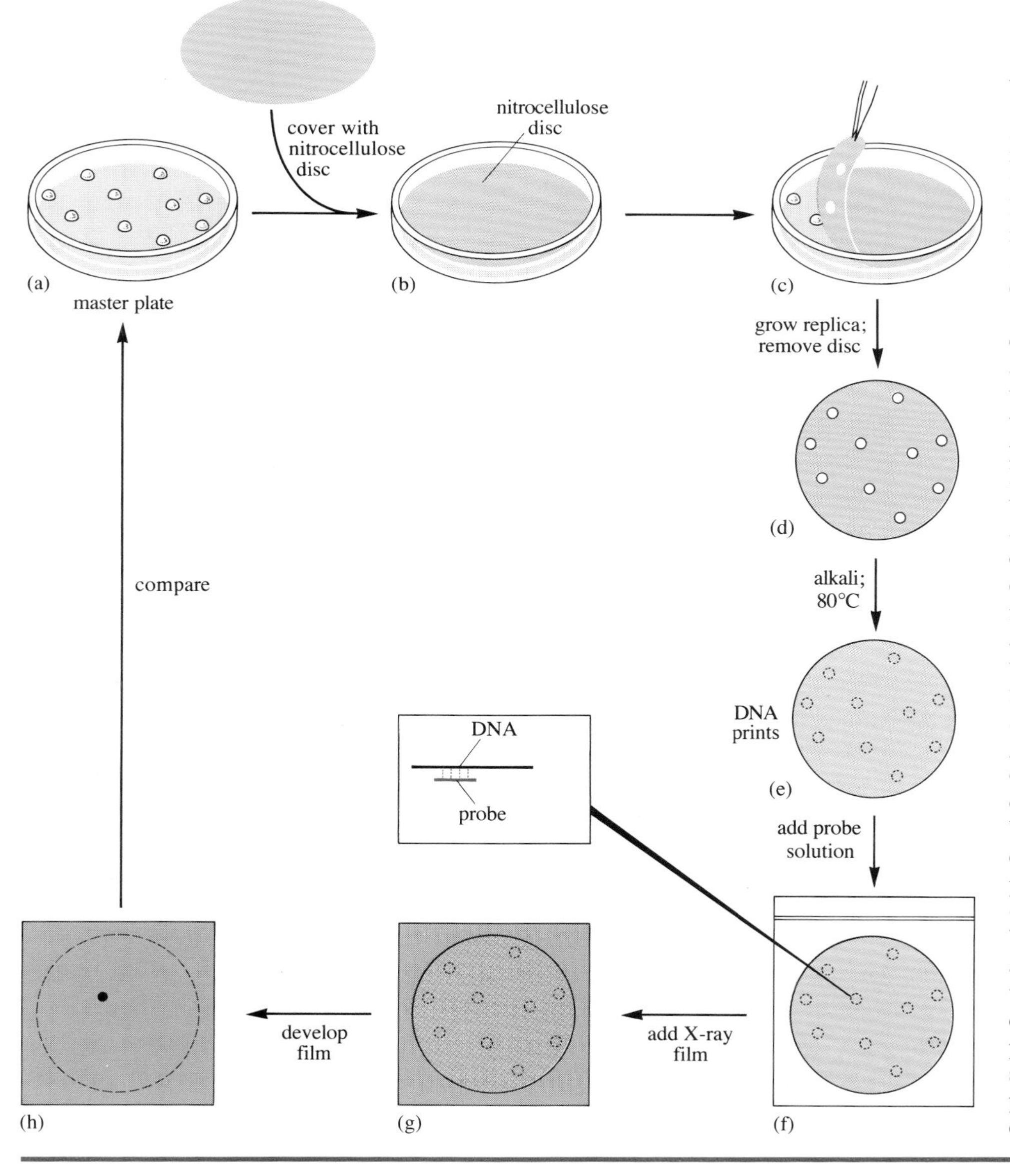

Figure 6.3 Probing for the gene. (a) The master plate, with colonies growing on the surface of the nutrient jelly. (b) A nitrocellulose filter disc (in pink) is pressed lightly against the surface of the master plate, thereby transferring a small portion of each growing colony onto the nitrocellulose disc. (c) When removed from the nutrient jelly, the nitrocellulose disc has an identical array of colonies to that on the master plate (d). It is thus an exact replica of the master plate. (These colonies, on the replica, can be allowed to grow further.) Treating the replica with alkali breaks open the cells in the colonies and separates the strands of their DNA. Protein is removed from the disc which is then baked at 80 °C, leaving DNA 'prints' of the colonies (e). The nitrocellulose filter disc is then immersed in a solution containing the radioactive gene probe (f). The probe binds only to single-stranded DNA derived from the desired gene (the binding is shown in the inset enlargment). The disc is then placed in contact with X-ray film (in grey) and left in the dark (g). When the film is developed, radioactive decay on the disc (i.e. within bound probe) shows up as a dark spot(s) on the film (h). Comparing the position of such a spot(s) with the original master plate indicates the colony (or colonies) containing the desired gene.

disc is then treated to remove protein but leaving the single-stranded DNA from each colony bound to the disc *at exactly the same sites as were the colonies themselves.* Finally, the disc is baked at 80°C, thus fixing the single-stranded DNA firmly in place (Figure 6.3e). As comparing (e) with (d) in Figure 6.3 will show, this leaves us with a nitrocellulose disc bearing the 'DNA prints' (also called 'DNA ghosts') of each of the colonies present on the replica and, hence, present still on the master plate (Figure 6.3a). It is these DNA prints that are now tested, *probed*, in our search for the desired gene.

6.4.2 Using gene probes

The production of DNA prints, at the same sites on the replica as the colonies that they come from, takes us as far as stage (e) in Figure 6.3. To understand the next stages, the actual probing of the DNA to find the desired gene, it is necessary to digress slightly. We need to consider again what a gene is. Remember we are trying to find one particular gene. So we must use some feature unique to just that gene.

▷ What feature is unique to *any* particular gene?

▶ The information that it contains. This information is encoded in the form of a base sequence; that is, a gene is a particular sequence of bases within a DNA molecule. Each gene has a base sequence unique to it alone (Box 4.1).

As a gene has a unique base sequence, it is the base sequence that we must exploit in our search for any particular gene. From the standpoint of its information, a gene can be regarded as a particular base sequence in *just one* of the two strands (the *coding strand*) of a double-helical DNA molecule. But, from a *physical* point of view, the base sequence of a gene is part of a double-stranded molecule. That is, the bases in the coding strand are paired with their complementary bases in the opposite (non-coding) strand of the double helix.

▷ What is meant by base complementarity?

▶ In a DNA double helix, each base in one strand of the helix is bound, via hydrogen bonds, to a base in the opposite strand. This binding is specific—adenine (A) pairs only with thymine (T), and cytosine (C) only with guanine (G). This specificity is referred to as *base complementarity* (Box 4.2).

Thus in an entire molecule of DNA, or a region within it (such as a gene), each strand is complementary and hence bound, as a whole, to the opposite strand. But, as you are aware (Box 4.2), hydrogen bonds are weak. Thus it is possible to break the hydrogen bonds between complementary base pairs relatively easily. For example, this can be done by adding strong alkali to a solution of DNA. Such treatment does not affect the much stronger covalent bonds, such as those linking nucleotide to adjacent nucleotide within each strand of DNA. The net result of such treatment is that the two strands of the DNA double helix remain individually intact but separate physically from each other. This leaves a preparation of **single-stranded DNA**. This is what happens when alkali is used to break open cells in colonies on a nitrocellulose disc. Hence, as stated earlier, the DNA prints fixed on the disc (Figure 6.3e) are of single-stranded DNA.

The critical difference between double-stranded DNA and single-stranded DNA is that in the latter the bases are *unpaired*. Thus, they are *free to pair* with other, complementary, bases. It is this that forms the basis of our probing for the desired gene.

What is done is to treat the baked disc bearing the DNA prints with a solution containing molecules of a specific *polynucleotide* (Figure 6.3f). This polynucleotide is *single-stranded*, a single chain. The individual monomers can be either deoxyribonucleotides (hence giving a poly*deoxyribo*nucleotide as in DNA; Figure 4.3) or ribonucleotides (hence giving a poly*ribo*nucleotide as in RNA; Box 5.1); it does not matter. What does matter is that the *sequence* of bases (within the nucleotides) in the polynucleotide chain is precisely complementary to the sequence of bases in the gene that we seek. Under defined conditions (temperature, concentration of ions), the polynucleotide will bind, via hydrogen bonds, to any *exactly complementary* sequence among the single-stranded DNA present on the baked disc, *and to that DNA alone*. As we use a polynucleotide that is precisely complementary in sequence to our desired gene, any such binding must indicate the presence of that desired gene (i.e. the presence of a single-stranded DNA print containing the base sequence of that gene). The polynucleotide in effect 'probes' the DNA on the disc, detecting and binding *only* to DNA complementary to it (if any such is present), i.e. only to DNA from our desired gene. Such a polynucleotide is called appropriately a **gene probe**.

In preparing a gene probe (of which more in Section 6.4.3), we 'label' it. This labelling is done most readily by making the polynucleotide radioactive, say by incorporating atoms of radioactive phosphorus (of the isotope ^{32}P) into it. By detecting the label, we locate the site of the binding of the probe and hence the location of the DNA print within which single-stranded DNA from our desired gene lies. We can easily detect the radioactive label by a technique known as **autoradiography**. In this we place a sheet of X-ray film in contact with the treated disc (Figure 6.3g), in the dark. Where radioactive decay occurs on the disc (i.e. within the bound probe) the emissions cause local deposition of silver grains in the emulsion and when the film is developed these appear as a dark spot (Figure 6.3h). (Incidentally, radioactive decay was first discovered in 1896 by its effect on photographic plates placed near to uranium compounds.) Comparing the position of the spot(s) on the film with the pattern of colonies on the master plate, identifies that colony or colonies (if any) that must contain the desired gene.

This technique of screening clones using gene probes is a general one. Any gene, indeed any sequence of DNA, can be detected, *always providing a suitable probe is available*. That is, to find a particular gene, we must have a *specific* gene probe, a polynucleotide with a base sequence complementary to that (in one or other of the two strands) of the gene we seek.

▷ Why does it not matter *which* of the two strands of the desired gene a gene probe is complementary to?

▶ The desired gene itself consists of two complementary strands (the coding and non-coding ones). These are separated on the nitrocellulose disc by alkali (Figure 6.3e) to form a single-stranded DNA print. A gene probe is complementary to just one of the two strands of the desired gene. But if the probe detects (binds to) such a strand, that could only have come from the gene (i.e. consisting of two strands) in the first place; so whichever strand is detected by the probe indicates the presence of the gene in the colony from which the DNA print derives.

A major problem is how to actually obtain a suitable specific gene probe.

6.4.3 Getting gene probes

Though the technique of gene probing is a powerful one, obtaining the necessary specific probes can present a significant problem. Consider again what a gene probe must be. It must be a polynucleotide whose base sequence is precisely complementary to that of the gene being sought. Yet in many, if not most, cases *we do not know the base sequence of the gene being sought in advance of actually obtaining that gene*. So obtaining a suitable probe can rely but rarely on knowledge of what its exact base sequence should be. So getting suitable probes often depends on methods where such knowledge is not needed. A detailed treatment of these methods is well beyond the scope of our present discussion but we will discuss very briefly the *principles* of a few such techniques. As these principles depend on some understanding of how genes code for proteins (polypeptides), it may be advisable first to look again at Box 5.1.

Getting the right message

By definition, a gene probe has a base sequence complementary to that (in one or other strand) of a particular gene.

▷ Considering the material in Box 5.1, can you suggest what could serve as a gene probe?

▶ The messenger RNA (mRNA) from a gene.

As an mRNA is transcribed from a gene (from one strand, the coding strand) by means of base complementarity, the mRNA must automatically be complementary to that gene. Thus, if we can obtain the mRNA corresponding to a desired gene and attach a suitable (say, radioactive) label to it, then we have an ideal gene probe; precisely complementary to, and hence specific for, our desired gene alone. The problem then is how to obtain such an mRNA.

As you know from an earlier context (cDNA cloning; Section 5.3.2, in particular), the place to look for the mRNA complementary to a particular gene is in cells known to make the protein corresponding to that mRNA/gene. So when we extract our mRNA (our probe) from some of the foreign cells, we must make sure that we choose ones that make the protein in question. Cells *abundant* in the particular protein (and hence in the corresponding mRNA, required as a probe) would be a good source. Thus, for example, if we wanted to probe bacterial clones for the (foreign) gene for human haemoglobin, using human red blood cells as a source of the mRNA, rather than, say, human brain cells, would be a good idea. But even cells abundant in one particular protein (and hence in the correponding mRNA) will contain a *mixture* of mRNAs. And to get a pure specific probe, complementary to just one gene, we need one particular mRNA. So, in essence, the problem revolves around how to separate mRNAs and identify the one we want. I shall duck the details of how this can be done—several methods exist. Suffice it to say, that in practice it is easier to separate the mRNAs not as mRNAs as such *but as cDNA copies*. Thus our final probe is not an mRNA as such but a cDNA copy of it. This cDNA can then be used to probe the bacterial clones to find the desired gene. (If you find this approach somewhat 'circular' and hence confusing—using a cDNA to probe for a gene—you are right! The reasons for this approach relate to technical practicalities about separating mRNAs. The only important point to remember is the principle that, in effect, we are using the mRNA of a gene as a probe for that gene.)

Knowing the polypeptide product

One alternative to isolating and purifying the mRNA corresponding to a particular gene is, *in effect*, to synthesize the mRNA; at least to synthesize a polynucleotide complementary to that gene. (As mentioned earlier in Section 6.4.2, in practice, this polynucleotide can be of ribonucleotides or deoxyribonucleotides—it is only the base sequence that matters.) It is possible to synthesize such polynucleotides by chemical means. We can chemically attach nucleotide to nucleotide to produce a polynucleotide probe complementary in base sequence to the gene we seek; in effect a 'synthetic mRNA'. But, just as we would not know the base sequence of the gene in advance, how would we know the base sequence of its complement, its mRNA? So how do we know what sequence to give this synthetic mRNA?

The first point is that the polynucleotide probe (the synthetic mRNA) need not be complementary to the entire gene. It is sufficient to have a probe that is much shorter than the gene, provided it is long enough to be complementary to (and hence bind to) a base sequence unique to that gene. Obviously very short probes would be no good. For example, a probe just four nucleotides long—say, AGGC—would be of little use, as sequences complementary to such short sequences are likely to occur many times, in many different genes, in the total genome of an organism. In practice, a polynucleotide some 15–20 nucleotides long would be big enough to be complementary to part of just one unique gene. Such short polynucleotides are usually called **oligonucleotides** (*oligo* means 'few'). That still leaves the question of how we know what sequence to make the oligonucleotide probe (in effect, a 'short synthetic mRNA'). We still need to know the appropriate base sequence for at least part (15–20 bases) of our synthetic mRNA. Once again, how can we know this in advance of cloning, isolating and studying the gene or its mRNA?

The answer relates to knowledge not of the gene itself but of its unique polypeptide product. Thus this approach will only work with *structural genes* (Box 4.1). This does *not* rely on expression of those genes in the bacterial clones, only on the fact that they can be expressed in principle (for example, in their native cells), i.e. that they do encode the structure of polypeptides. The technique also requires us to know something of the structure of the polypeptide product, in particular something of its amino acid sequence.

Let us assume that we know a stretch of five or six amino acids within the polypeptide chain. We will assume further that such a sequence will occur only in this polypeptide and no other—the amino acid sequence is unique to this polypeptide. That unique sequence of amino acids must be encoded by a similarly unique sequence of bases in DNA, a sequence of bases that forms part of the gene for that polypeptide. So, knowing a unique sequence of amino acids within the polypeptide, we can deduce the corresponding unique base sequence within the gene and the unique sequence in the mRNA. Thus we can deduce the required base sequence for an appropriate oligonucleotide probe (a short synthetic mRNA).

Deducing a base sequence for a short synthetic mRNA from a known amino acid sequence within the polypeptide depends on knowing the genetic code—the correspondence between particular codons and amino acids (Table 5.1). That is, we can 'work backwards'—deducing what sequence of bases must give rise to a particular amino acid sequence. *But* as a look at Table 5.1 will show, it will not in general be possible to deduce just one possible base sequence. This is because the code is a *degenerate* one; that is, most amino acids may be coded for by more than one codon. Let us illustrate the problem for a sequence of just three amino acids.

For example, say we know that the following sequence of three amino acids occurs in the polypeptide in question: Met Lys Ileu.

▷ What would be the base sequence in the portion of mRNA corresponding to those amino acids? You will need to refer to Table 5.1.

▶ Though there is just one codon for Met (AUG), there are two for Lys and three for Ileu. It is therefore *not* possible to give an unambiguous base sequence for the corresponding portion of mRNA. But alternative sequences can be deduced. These are: AUG AAA/G AUU/C/A where alternative bases are as indicated. Thus there are six possible sequences deducible from the known sequence of just these three amino acids.

In trying to derive the base sequence suitable for a short synthetic mRNA (probe) from five or six amino acids (corresponding to an mRNA of 15 or 18 base's, respectively; as the genetic code is a triplet one), we might come up with, say, 30 or 40 different possible base sequences. Yet, though, *in principle*, any one of these could code for the sequence of amino acids, only one will *in fact*. That is, the gene we seek will have a base sequence complementary to only one of the theoretically deducible mRNA sequences. So, only a short synthetic mRNA with the one correct sequence will bind to the gene we seek. How do we know which of the many (say, 30 or 40) theoretical sequences is the right one, which oligonucleotide to synthesize?

The answer is that we don't know! So we synthesize a mixture of all the possible theoretically correct oligonucleotides (radioactively labelling them also) and use the mixture as our short synthetic mRNA probe. We know that within that mixture will be one oligonucleotide precisely complementary to (and hence that will bind to) part of the gene we seek. The other oligonucleotides in the mixture are irrelevant, they will not bind to anything (they could only be complementary to an 'equivalent to our gene'—a slightly different gene that codes for the same amino acid sequence, a gene that we may assume does not exist in that cell). Binding of the one correct oligonucleotide (detected by its radioactive label) to a DNA print (Figure 6.3) signals the presence of the base sequence complementary to that oligonucleotide in that DNA. That is, it signals that the print must have DNA derived from the desired gene.

Using related genes

The final method that we shall discuss involves using genes themselves as probes.

Let us assume that we have already cloned and located a particular foreign gene (i.e. we know which clone of bacteria it is in). Then it is possible to isolate the recombinant plasmids from just that clone (they are all the same and contain the same foreign gene). Without detailing how, it is possible to remove a foreign gene from a plasmid (we 'clip it out'—cut out part, or all, of the insert in the plasmid using appropriately chosen restriction enzymes). This gene can then be used as a gene probe.

What is done is to label the DNA of the foreign gene (say, radioactively) and separate the two strands of the double helix.

▷ How might this be done?

▶ With strong alkali.

A solution of this radioactive single-stranded DNA can then be used as a gene probe (i.e. used in step (f) in Figure 6.3). But what would it probe?

- Firstly, it might be used to find the *same* gene in a different collection of clones. That is, having cloned and isolated the gene once it can be used as a probe for the same gene cloned in a separate experiment.
- Secondly, it might be used to probe for *related* genes. For example, a gene cloned and hence isolated from one species of organism might be used as a probe (for the 'same' gene) in another collection of clones derived from a different but related species. Generally, there will be some similarity in the base sequence of the same gene from related species (say, the gene for human insulin and that from a pig or cow) but also some differences. So a probe derived from the gene (i.e. single-stranded DNA) of one species might bind to a DNA print from another, provided

that these genes were closely related. But we argued earlier that gene probes are designed to be as specific as possible, just binding to one (complementary) base sequence and that alone — hence identifying just *one* desired gene from among many possible ones (Figure 6.3). Under the conditions (such as the temperature and concentration of ions) chosen earlier (Section 6.4.2) absolute precision is required — a complete complementarity of bases; so indeed only one unique gene is located. However, by juggling the conditions we can make that binding less precise — a few mismatched bases (non-complementary and hence unpaired) can be tolerated and hence one gene can be used as a probe to pick out a related, but slightly different, one; as, say, from a different species.

6.4.4 Other probes, other uses

Though several methods exist for producing probes, doing so is not always easy. For the methods outlined above, we either need a way of obtaining and purifying a particular mRNA (or its cDNA equivalent) or knowledge of the amino acid sequence of a protein product or possession of an earlier isolate of the gene (or its close relative). Where such material or information is lacking, even less direct methods of making probes must be sought. Yet gene probes are powerful tools and so the incentive to find ways of producing them is often considerable.

Gene probes provide the advantage that a gene can be detected in a clone even if it is not expressed there. Thus such a technique is more general than one depending on detecting a protein product (Section 6.3). So probing for a foreign gene directly is frequently the starting point for identifying a clone. This may be true even if the gene sought is expressible in principle.

▷ What would render a foreign gene in bacteria 'expressible in principle'?

▶ There are three criteria that we have mentioned that it would need to fulfil. First, it is a structural gene and therefore encodes a polypeptide product (Box 4.1). Secondly, it is not a split gene (Box 5.2), or, if split in its 'native' state, the introns have in effect been excised via the cDNA route. Finally, the gene has an associated promoter compatible with the host cell (*E. coli*) RNA polymerase (Sections 5.1 and 5.3.1).

Once located in its clone, such a gene can then be subjected to further 'tinkering' (which I need not discuss, but which you will realize is possible using the repertoire of cutting and splicing techniques), e.g. putting it under the control of powerful and compatible promoters so that it will be expressed efficiently, determining its base sequence, or whatever.

So far we have discussed gene probes only in the context of identifying which clone of bacteria carries a desired foreign gene. But, in principle, a gene probe can be used to detect a foreign gene *wherever* it is — in a bacterial clone, inside another (non-bacterial) host cell (Chapters 7 and 8), even inside its 'native' cell (i.e. where, in fact, it is not a 'foreign' gene).

The general applicability of the technique of gene probing — for example, irrespective of where the gene is, independent of expression of a gene, usable even if the gene is not a structural one (perhaps a control gene; Box 4.1) — means also that gene probes can be used to detect the presence or absence of a gene *even if we do not know its precise function*. The problem is to find a method of making the probe.

Given the power of the approach and its generality, it is not surprising that gene probes have been applied widely.

Most excitingly, they have been of great use in helping identify genes and detecting minor differences between different alleles of the same gene (Box 2.1). This has given them a recent use in the diagnosis and investigation of inherited diseases—such as muscular dystrophy and cystic fibrosis—a topic dealt with in Section 9.2.2.

6.5 From shotgun to rifle

There is no doubt that a shotgun approach can work. Screening a large number of clones may prove tedious, but with a complete gene library we at least know that our desired gene should be there somewhere (and, hopefully, intact). Using a cDNA library is somewhat more precise and cuts down the number of clones to be screened.

However, in some circumstances where we require the product of a foreign gene to be produced in bacteria we can by-pass the needle-in-a-haystack aspects of screening altogether—convert 'shotgun' to 'rifle'. This does not involve transferring a large number of different gene-size fragments of DNA (Figure 4.2) or cDNAs (Figure 5.6). Instead we transfer a single type of DNA alone, that of the 'gene' we are after. In fact we do not transfer an actual gene isolated from a foreign cell; instead we transfer a **synthetic gene**.

What is done is to synthesize chemically molecules of a DNA that has a base sequence the same as the natural gene or *equivalent to it*. Then, using techniques like those outlined in some later stages of Figures 4.2 (e and f) and 5.6 (c and d), we can splice these synthetic genes into plasmids and obtain bacteria that contain recombinant plasmids, all of which carry the same inserted synthetic gene. Assuming an appropriate expression vector is used, the synthetic gene will be expressed to give the required protein product. Other than to check that transfer has occurred as planned, there is no need to screen clones to see which contains the desired (synthetic) gene—only one type of gene is involved. This is obviously a precise technique. The problem lies in synthesizing the gene. This problem is twofold.

1 To synthesize a gene we must obviously know its base sequence. Sometimes this information can be obtained by taking the *natural* gene and determining its base sequence using analytical techniques that have been developed over the last 15 years or so. However, this would mean first *isolating* that gene and this might require shotgun cloning techniques, the very thing we might be trying to avoid!

▷ Can you think of an alternative method of determining the base sequence of a gene, assuming that we know the amino acid sequence of its polypeptide product?

▶ Using the known codon assignments in the genetic code (Table 5.1) we can 'work backwards' on paper and deduce a sequence of bases for the corresponding gene.

▷ Is it possible to deduce an unambiguous base sequence for a gene working backwards from the known amino acid sequence of the polypeptide product?

▶ No, as already seen in a different context (Section 6.4.3). As a glance at Table 5.1 will show, with two exceptions (Trp and Met) there is more than one codon for each amino acid. (The table shows mRNA codons but obviously each of these has a corresponding, complementary, DNA codon; Box 5.1).

Thus, all we can hope to do by working backwards from an amino acid sequence is to deduce a base sequence for a gene that *could* give rise to that polypeptide; not necessarily the sequence actually in the *natural* gene for that polypeptide but one with *equivalent* codons (i.e. codons coding for the same amino acids). And this may be all we have to do—a synthetic gene with our deduced sequence should code for the same polypeptide product as the natural gene and it is the polypeptide product we are after. That is why we wrote 'base sequence the same as the natural gene or *equivalent to it*'.

Also, working backwards from the polypeptide sequence would only give us the actual coding sequences, the *exons*, not the sequence of any introns that might exist in the natural gene. This is of course sufficient, as we need to synthesize a 'gene' with just the exons and excluding any introns that might exist in the natural gene.

▷ Why would we not want any introns in our synthetic gene?

▶ We wish to obtain the polypeptide product of the synthetic gene and any introns present would prevent this gene being expressed in bacteria to give the product we seek. After all, we have to go to some considerable lengths to by-pass introns when using 'natural genes', using instead cDNAs derived from natural mRNAs (Section 5.3).

2 We need techniques for synthesizing specific DNA molecules, molecules with defined base sequences. This means chemically joining together nucleotide bases in the correct sequence. This was extremely difficult even a dozen years ago, and the longer the DNA molecule the harder it was—a synthetic gene for somatostatin, a small hormone containing just 14 amino acids, being a triumph in its day; as was the synthesis of synthetic genes for the two chains of another hormone, human insulin, some 21 and 30 amino acids long.

▷ How many base pairs would each of these synthetic genes have to contain?

▶ Remember that each codon is a triplet of bases (Table 5.1). So the synthetic gene for somatostatin must comprise $3 \times 14 = 42$ base pairs; those for the insulin chains, 63 and 90 base pairs.

However, techniques continued to improve—for example, the gene for human α-interferon, over 500 base pairs long, has been synthesized. (Thus, a third approach to producing human α-interferon is available. But this depends on considerable detailed knowledge of the gene, or the protein, in order to be able to synthesize an 'equivalent' gene sequence; Activity 6.1.) These days, given the right information (see above), we can use highly automated 'gene machines' to synthesize quite large genes with great accuracy and fairly rapidly.

However such information on base sequence may not be easy to obtain. The use of synthetic genes is still a relative rarity. Despite their apparent relative crudity, shotgun techniques can prove simpler in the long run. We can, however, use the classic experiments on human somatostatin to illustrate the use of synthetic genes. It will also serve as a way of recapitulating and reinforcing some of the other aspects of genetic engineering in bacteria that we have dealt with in these last three chapters.

6.5.1 Somatostatin — a first for synthetic genes

Somatostatin is a hormone produced by the hypothalamus (a small structure at the base of the brain). It has various biological effects including inhibition of the release of other hormones such as insulin and glucagon. Chemically, somatostatin is a peptide (i.e. a small polypeptide) comprising a single chain of just 14 amino acids; the amino acid sequence of this chain has been known for around 20 years. In 1977 Itakura and his colleagues used the known amino acid sequence to deduce a possible base sequence for a synthetic somatostatin gene. They then chemically synthesized this gene. The base sequence of this synthetic gene is shown in Figure 6.4.

		1	2	3	4	5	6	7	8	9	10	11	12	13	14			
	Met	Ala	Gly	Cys	Lys	Asn	Phe	Phe	Trp	Lys	Thr	Phe	Thr	Ser	Cys	Stop	Stop	
AATTC	ATG	GCT	GGT	TGT	AAG	AAC	TTC	TTT	TGG	AAG	ACT	TTC	ACT	TCG	TGT	TGA	TAG	
G	TAC	CGA	CCA	ACA	TTC	TTG	AAG	AAA	ACC	TTC	TGA	AAG	TGA	AGC	ACA	ACT	ATC	CTAG

Figure 6.4 The base sequence of the synthetic gene for somatostatin and the corresponding amino acids coded for. The coding strand is in red. In the actual somatostatin, the cysteines at positions 3 and 14 react to give a disulphide bridge (Box 5.1; and see Figure 6.5, later). (In the diagram the bases have been arranged in triplets to make their correspondence with the amino acids easier to read.)

▷ The coding strand is the lower of the two shown in Figure 6.4, where it is coloured red. Using the data in Table 5.1 confirm that indeed this is the coding strand. (Assume this strand is transcribed to give mRNA reading from left to right and that transcription starts at the first complete triplet of bases. You do not need to examine the whole of the message; consider, say, the first five or six codons.)

▶ The coding strand must be transcribed to give mRNA using the rules of base complementarity (Box 5.1). Thus reading from left to right, from TAC (the first complete triplet), the complementary mRNA would have the following base sequence:

AUG GCU GGU UGU AAG AAC UUC UUU UGG... etc.

Comparing this mRNA sequence with the codons given in Table 5.1, this would code for the following amino acid sequence:

Met Ala Gly Cys Lys Asn Phe Phe Trp... etc.

This is in fact the actual sequence of amino acids shown in Figure 6.4 and so this indeed confirms that the lower strand of the DNA of the synthetic gene is the coding strand.

There are several other points to note about this synthetic gene:

- Somatostatin contains 14 amino acids yet the base sequence shown codes for 15, starting with Met. Somatostatin starts at the Ala.
- The DNA contains *two* triplets in sequence, ACT and ATC, that correspond to, complementary, stop codons in mRNA (UGA and UAG respectively; Table 5.1). These will ensure that the mRNA from this synthetic gene will say 'end of message' when these are reached by the ribosomes (Box 5.1). Having two presumably provides 'insurance' in this experimental situation in case the first one is missed.
- The synthetic gene has no adjacent promoter.
- There are a few additional bases either side of the coding region; some of these are unpaired (i.e. only one or other strand of the DNA helix is present, giving 'ragged ends' to the molecule). These need not concern us, except to say that they were important in helping splice this synthetic gene into plasmids (Section 4.4).

Itakura and colleagues spliced this synthetic gene into an opened plasmid circle that had already been specially constructed so as to contain an *E. coli* gene, that for an enzyme called β-galactosidase, together with its associated promoter region.

▷ What is meant by a plasmid being 'specially constructed'?

▶ A naturally occurring plasmid modified by the standard cutting and splicing methods (Chapter 4) so as to remove or insert other DNA. In this case, the gene for β-galactosidase plus its neighbouring promoter, which also occur normally on the *E. coli chromosome* (Figure 4.1a), had been spliced into a plasmid.

The synthetic somatostatin gene was spliced into the plasmid *within* the β-galactosidase gene, near to its far end (i.e. the end corresponding to the C-terminal portion of β-galactosidase; Box 5.1); as shown in Figure 6.5a.

When taken up by *E. coli* the cells produced a hybrid polypeptide (Figure 6.5d).

▷ Why is a hybrid polypeptide produced?

▶ Splicing the synthetic somatostatin gene within that for β-galactosidase means that the whole base sequence (from the N-terminus of β-galactosidase up to and through the somatostatin gene) is *under the control of the β-galactosidase promoter*. Thus when mRNA synthesis is initiated by this promoter (Figure 5.3) a single hybrid mRNA is produced which when translated yields a hybrid polypeptide (Figure 5.7). The specially constructed plasmid using β-galactosidase in this way acts as an *expression vector* (Section 5.3.1). Being native to *E. coli*, the β-galactosidase promoter is naturally compatible in the *E. coli* cells which take up the plasmid.

Finally, Itakura and colleagues needed to release the somatostatin in the hybrid polypeptide from the long β-galactosidase portion to which it was covalently bound (Figure 6.5e). This was achieved by treating the hybrid polypeptide with *cyanogen bromide*. This is a chemical that specifically reacts with methionine (Met). If you look back at Figure 6.4, you will see that somatostatin *proper* contains none of this amino acid. But Met is the additional amino acid encoded at the very start of the synthetic gene. Thus when the hybrid polypeptide is treated with cyanogen bromide the somatostatin portion proper is unaffected but the additional Met is affected and this results in the hybrid polypeptide chain being cleaved at this point. This has the desired effect of releasing precisely intact somatostatin from the hybrid polypeptide (Figure 6.5e). And that was the reason for adding the bases coding for Met in the first place (Figure 6.4); planning ahead as to how to release the correct portion from the hybrid—a cunning piece of experimental design.

Cyanogen bromide also breaks up the β-galactosidase portion into fragments, but this is of no consequence for the experiment.

▷ What, however, does the fragmentation of the β-galactosidase portion of the hybrid tell us about its amino acid sequence?

▶ That it contains one or more Met residues.

Incidentally, Itakura and colleagues used an antibody assay to detect the released somatostatin (Section 6.3).

Activity 6.2 *You should spend up to 20 minutes on this activity.*

Imagine that a molecule somewhat like somatostatin has just been isolated from a rare variety of Peruvian llama and given the name *llamastatin*. Llamastatin is a single peptide chain with 14 amino acids identical to those in somatostatin; except that at positions 6 and 14 from the N-terminus, amino acids other than phenylalanine (Phe) and cysteine (Cys) occur. From the amino acid sequence of llamastatin, a possible base sequence is deduced and the gene synthesized. Additional bases are included to provide for two stop codons and a Met before the normal N-terminal Ala, plus bases to assist in splicing the gene into plasmids. The base sequence of this synthetic DNA is identical to that produced by Itakura and colleagues for somatostatin except for different bases at positions corresponding to amino acids 6 and 14 in somatostatin. These have been altered to take account of the differences in llamastatin at these positions. The new triplets in the coding strand are TAC and GAA, at positions corresponding to amino acids 6 and 14, respectively.

A technique identical to that used by Itakura and colleagues is then used to clone the synthetic gene and produce a hybrid polypeptide product. However, following treatment of the hybrid product with cyanogen bromide no llamastatin can be detected, despite using an antibody specifically prepared to bind to it. Similarly, an assay for the known biological activity of llamastatin fails to detect any sign of it. Can you suggest what might have gone wrong? (You will need to examine Figures 6.4 and 6.5 and Table 5.1; remember that the table gives *mRNA* codons.)

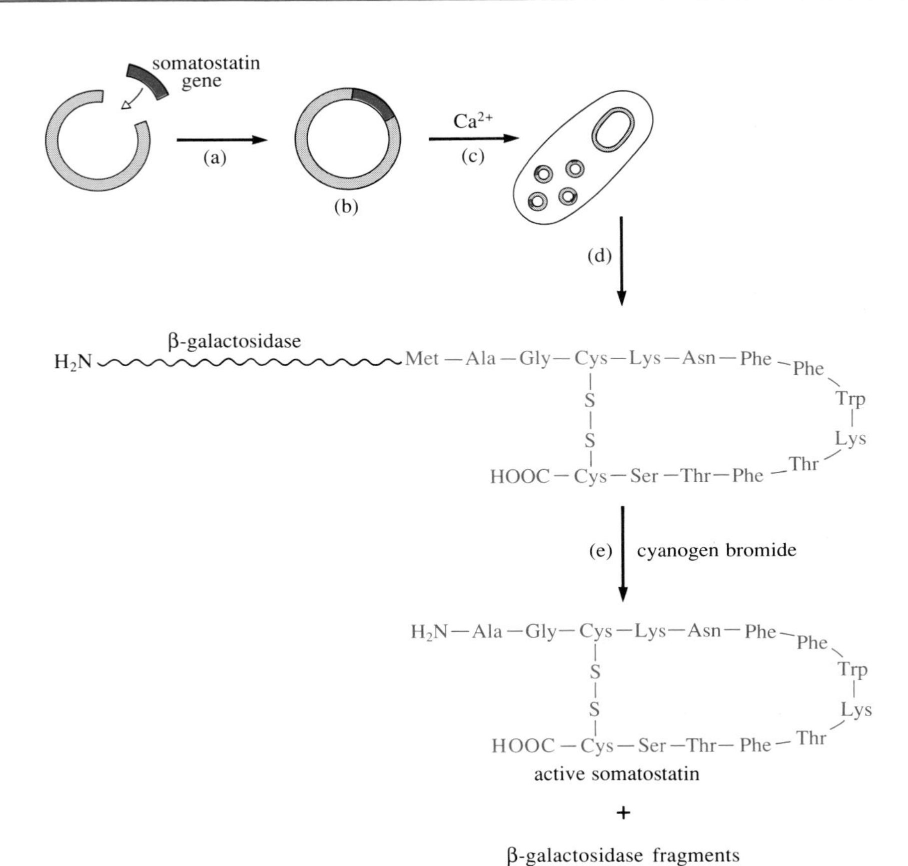

Figure 6.5 Producing somatostatin in *E. coli* using a synthetic gene. (a) The synthetic gene for somatostatin (in red) is spliced into a plasmid within a gene for β-galactosidase (which acts an expression vector) producing (b) a recombinant plasmid. (c) The recombinant plasmid is taken up by *E. coli* cells which (d) express the hybrid β-galactosidase–somatostatin gene to give a hybrid polypeptide. (e) When treated with cyanogen bromide, the hybrid polypeptide is broken, yielding fragments of β-galactosidase and intact somatostatin. *Note*: the disulphide bridge (—S—S—) in somatostatin forms after translation.

Summary of Chapter 6

1 Techniques whereby gene-size pieces of foreign DNA (generated by restriction enzymes) or mixtures of cDNAs are cloned are known as *shotgun techniques*. They result in the formation of *gene* or *cDNA libraries*. The problem is then to find the right 'book', the clone with the desired foreign gene or cDNA. This is done by *screening* individual clones. We can calculate how many clones will probably need to be screened.

2 Individual clones can be grown by separating individual cells by spreading them onto the surface of nutrient jelly; a technique called *plating*. Each cell grows and divides where it sits to produce a clone of its descendants visible as a mound or *colony*.

3 If the transferred foreign gene is expressed, yielding a protein product, the colonies containing this gene can be located by *screening for the product*, the protein.

4 If the protein product is an enzyme it may be convenient to detect it by using a suitable *specific enzyme assay*. Where no such suitable assay exists, or the protein is not an enzyme, it may be detected by using a *specific antibody* prepared against that protein.

5 Alternatively, the gene itself can be detected directly by means of a *gene probe*. This is particularly useful where the gene is not expressed. A gene probe is a *polynucleotide complementary* in base sequence *to* (one or other of the two strands of) *the gene sought*. The gene probe *binds specifically to DNA* extracted from the clone containing the *desired gene*.

6 Clones can be probed *en masse* by making a *replica* of the colonies on *nitrocellulose* and exposing the DNA *in situ* giving *single-stranded DNA prints*.

7 Several techniques exist for producing probes. For example, mRNA (or its cDNA equivalent), short 'synthetic mRNAs' (i.e. *oligonucleotides*), or related genes can all be used as gene probes.

8 In some cases the shotgun approach can be avoided by chemically producing a *synthetic gene* and transferring this into bacteria. To chemically synthesize a gene we must know its *base sequence*, something possible if the gene has already been isolated and its *base sequence analysed*. Or if the *amino acid sequence of its polypeptide product* is known we can *deduce* the sequence of the gene or an equivalent one.

9 *Somatostatin* was an early example of a foreign peptide produced in bacteria by transfer of a synthetic gene. The initial polypeptide produced was a hybrid, somatostatin being released by treatment of this with *cyanogen bromide*.

Question 6.1 Which of the items shown below would be necessary when: (a) cloning genes from DNA extracted from eukaryote cells, in bacterial host cells; (b) cloning genes via cytoplasmic mRNA extracted from eukaryote cells, in bacterial host cells?

(i) a suitable restriction enzyme

(ii) a medium containing calcium ions

(iii) reverse transcriptase

(iv) ligase.

Question 6.2 Look at Question 6.1 and assume that you also wish to get the foreign protein products produced in the bacterial host cells.

Assuming that the protein product of a particular gene/cDNA transferred was an enzyme, which of the following techniques could be used to detect the presence of a clone that produced that product?

(i) an assay specific for the catalytic activity of the enzyme

(ii) an antibody that binds specifically to that enzyme.

Question 6.3 Which of the following statements are true and which are false?

(a) A cDNA from a prokaryote cell will be shorter than the structural gene it represents.

(b) On average, the frequency of occurrence of each *different* clone (i.e. containing a *different* cDNA) in a cDNA library is equal.

7 From bacteria to other cells

We have considered the transfer of genes into bacteria in some detail, using the protocol in Figure 4.2 as our baseline. As we mentioned at the outset, this protocol is essentially one developed for *E. coli*, though with variations it can apply to other bacteria, to eukaryote microbes and to animal and plant cells too, as you will see shortly. We have already seen some variations on the theme, so it is worth summarizing what the key components of gene transfer into *E. coli* actually are.

Firstly, we need a method of obtaining the gene to be transferred.

▷ Can you identify three such methods?

▶ Cleaving foreign DNA with a restriction enzyme to give gene-size fragments; producing cDNAs from foreign mRNAs; chemically synthesizing the gene.

Thus you will note that though we generally speak simply of 'gene transfer', irrespective of the method used, in fact only in the first method is the *natural* gene as such involved. The other two methods do the 'equivalent' of using a cDNA or a synthetic gene.

▷ In what way do the first two methods differ from the last?

▶ The first two methods are 'shotgun' techniques by which a large number of different genes/cDNAs are cloned and the desired clone found afterwards by screening. The last method is more precise, a DNA with the same or equivalent base sequence to that of the desired gene is synthesized and it alone is transferred.

Secondly, we need a means of getting the foreign gene into a host cell, here *E. coli*. In the cases dealt with so far, this involves splicing the foreign gene into a *vector*, a plasmid, and allowing the cell to take up the recombinant plasmid. This it does in the presence of calcium ions.

Finally, assuming, as we have, that it is the protein product of the foreign gene that we are after, we need to get expression of that gene in its new, *E. coli*, surroundings. This may involve ensuring no introns are present, something that can be done by using a cDNA rather than a gene-size DNA fragment in the first place. It will also involve ensuring that a compatible promoter is appropriately located before the gene, something that may mean supplying one in the plasmid, the vector, itself (an expression vector).

In addition to these basic elements, there may well be other factors that have to be taken into account. This is particularly true if we wish to obtain the protein product in large amounts, say on an industrial scale. For example, we shall probably wish to maximize the output of the foreign protein *per cell*.

▷ There are various things we might take into consideration in trying to maximize output of the foreign protein per cell. In particular, suggest how we might maximize the level of mRNA (for the foreign protein) that is produced in the bacterial cell.

▸ In general, the more mRNA for a particular protein that is present, the more protein that will be made by translating it. There are two ways in which we might maximize the level of mRNA for the foreign protein.

1 Maximize the number of copies of the foreign gene from which the mRNA is transcribed—this can be done by using a plasmid that replicates well inside *E. coli* and so gives many copies of the plasmid per cell (Figure 4.1a). And many copies of a *recombinant* plasmid means many copies of the foreign gene that it bears.

2 Ensure that a powerful compatible promoter is present in front of the foreign gene; that is, a promoter that is recognized by the host cell RNA polymerase and efficiently so.

There are also things that we can do to ensure efficient *translation* of the mRNA that is produced. We need not consider these here, but, crucially, they can, to some extent, be manipulated by the experimenter.

Over the last 15 years or so our ability to tailor recombinant plasmids that achieve high yields of foreign proteins has developed considerably. So much so, that production of a foreign protein can virtually dominate the total protein production capacity of the cell, in some cases leading to a cell so 'constipated' with the foreign protein that it even precipitates or crystallizes inside the cell (Figure 7.1). Sometimes this allows highly efficient production of the desired protein but on other occasions recovering normal, *undenatured*, protein from such clumps may be difficult.

▹ **Undenatured** protein means protein in its normal state, with its precise three-dimensional shape intact. Why would it be important to obtain, ultimately, protein like this?

▸ In many, if not the vast majority, of cases we desire the foreign protein for some particular biological activity that it possesses. For example, insulin (Section 3.4) is desired for its activity as a hormone. The biological activity of a protein depends on its three-dimensional shape. In denatured proteins the normal shape is disrupted and so such proteins would most likely lack the very activity that we would be after. It is therefore important to obtain them *undenatured*.

Trial and error is always likely to be an important part of finding the ideal conditions for any particular foreign protein.

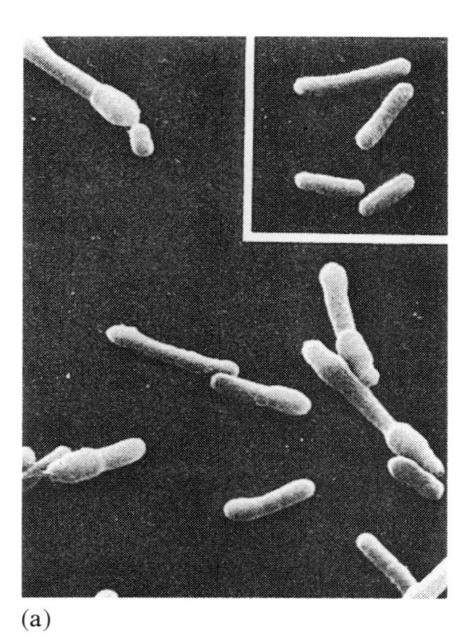
(a)

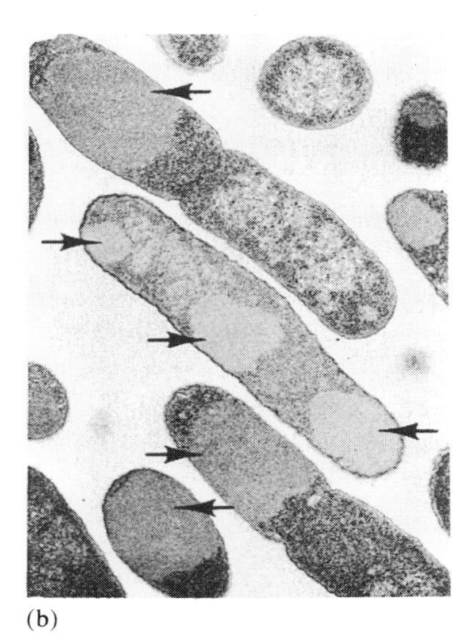
(b)

Figure 7.1 High yields of foreign protein in *E. coli*. (a) A scanning electron micrograph at × 5 300 magnification shows the surface features of the cells. Those cells producing the foreign protein (an expression vector hybrid polypeptide containing the human pro-insulin chain) positively bulge with it. Normal *E. coli*, shown inset, present more svelte outlines. (b) An electron micrograph (× 30 000) of thin sections of *E. coli* producing high yields of foreign protein reveals its presence as large aggregates (arrowed).

Thus, with these variations and others (which we need not consider) on the theme of how to obtain the foreign gene to start with and how to maximize its expression in novel surroundings to end with, the plasmid-led transfer outlined in Figure 4.2 does indeed represent a fundamental protocol. Likewise, the actual plasmids used in *E. coli* can vary widely and few that are used routinely are naturally occurring. Rather they are artificial constructs, derived from naturally occurring plasmids, designed to have features, such as good promoters, suitable for the particular job in hand.

▷ How in general can such artificially constructed plasmids be produced?

▶ By using the same sorts of techniques for cutting and splicing DNA that are employed later to insert foreign genes into plasmids for transfer into bacteria (Figure 4.2). Naturally occurring plasmids can thus have DNA sequences removed and others added, as desired, to create suitable artificial plasmids. The plasmid used in the somatostatin experiment (Figure 6.5) was one such artificial construct; in particular, the gene for β-galactosidase had been inserted to produce a plasmid that was an efficient expression vector.

But plasmids are not the only sorts of vectors available for gene transfer into *E. coli* and we need to discuss briefly some others.

7.1 Putting phage to work

As mentioned earlier in a different context (Section 4.3), bacteria can be infected by viruses known collectively as *(bacterio)phages*. In such infection, once inside a cell the genome of a single phage 'hijacks' the cell machinery causing the cell to make proteins coded for by *phage* genes and to replicate *phage* DNA. Ultimately, the newly synthesized phage proteins and molecules of phage DNA assemble together to give many new phages. These then rupture the cell, freeing them to invade and infect many more surrounding cells, and so on. One such pest for *E. coli* is a phage known as **bacteriophage λ** or more simply, and commonly, just **λ** (λ is the Greek letter *lambda*). The *infectious cycle* is shown diagrammatically in Figure 7.2; a much-enlarged λ is shown in Figure 7.3.

What is a pest for *E. coli* can prove a boon for genetic engineers. We are able to construct very useful vectors from λ. Put simply, we can take DNA derived from λ and insert foreign DNA into it, using once again the basic cutting and splicing tools of restriction enzymes and ligase (Sections 4.3 and 4.4). If this DNA is then mixed with λ proteins *in vitro* whole phage can assemble. Of course, in such phage the DNA is **recombinant DNA**; that is, a single molecule of DNA which is a splice of λ DNA and foreign DNA (i.e. *analogous* to DNA in recombinant plasmids). Such phage can thus be said to be *recombinant* phage. Infecting *E. coli* with recombinant phage leads to rapid production of many more such phage from each cell, via a cycle like that in Figure 7.2. All the progeny of such phages will carry their foreign DNA insert. The progeny phage can infect further cells (Figure 7.2e) and so on. Clones of

Figure 7.2 The bacteriophage λ infectious cycle in *E. coli*. (a) λ adheres to the surface of an *E. coli*, 'bores' a hole in the cell wall and (b) injects its DNA into the cell. The λ genes then 'program' the cell resulting in (c) production of λ proteins and replication of λ DNA. (d) λ DNA and proteins assemble to give many λ phages. (e) The cell bursts and the λ are then free to infect other cells. (*Note*: for simplicity, the *E. coli* chromosome is not shown.)

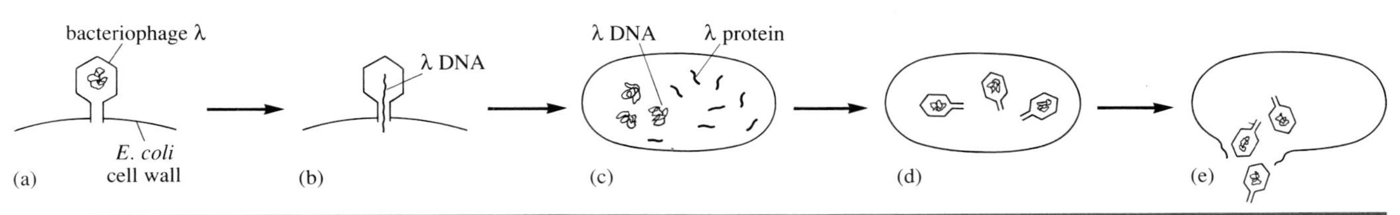

phage result. Once again, if the source of the foreign DNA has been of a 'shotgun type'—that is, either from restriction enzyme cleavage of large molecules of foreign DNA *or* as a mixture of cDNAs—the resulting phage must be screened to find those carrying the desired gene (we need not consider how the screening is done). The difference is that the desired gene will be present in a clone of phages, rather than in a clone of bacterial cells. But, as may have occurred to you, what advantages, if any, has using phage over using plasmids?

You may recall that plasmids can only tolerate relatively small inserts of foreign DNA (Section 4.3); large ones tend to be lost. Where we want to insert a single prokaryote gene or 'a eukaryote gene without introns' (i.e. a cDNA) this limited capacity is probably adequate, i.e. beyond the average length of DNA we need to insert. This is what we would want normally where expression of the foreign gene is the main objective (Chapter 5).

But say we want a whole eukaryote gene complete with introns? This is unlikely to be our aim if we want expression of a gene in *E. coli*. But there are circumstances where we might need the whole gene, introns and all. For example, we may wish to study the detailed structure of a gene, i.e. determine its *complete* base sequence. Then we might need a vector that can accomodate large lengths of DNA. For whereas a gene, or cDNA, without introns may well be just hundreds or a few thousand base pairs long, an intact one with introns can be very much longer. For example, consider one polypeptide chain of the enzyme uricase from soya beans. The coding portions, the exons (Box 5.2), of the gene for this polypeptide total a mere 300 base pairs. The whole gene, complete with its 7 introns, runs to 4 800 base pairs. Indeed much longer genes exist.

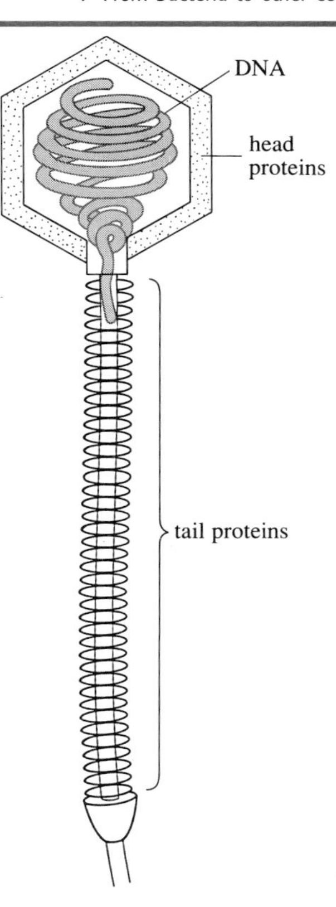

Figure 7.3 Schematic drawing of bacteriophage λ. A good deal is known about λ, including the complete sequence of bases in its genome—a single molecule of DNA 48 502 base pairs long. A considerable portion of this DNA codes for functions that are not needed for the infectious cycle or for packaging DNA plus proteins to give particles of intact phage.

λ too has its limits as to how much DNA it can accomodate. There is a limit to the quantity of DNA that can be packed into the assembled virus. But by deleting portions of the λ genome that are not needed for assembling DNA into whole phage or for the infectious cycle (Figures 7.2 and 7.3), we can produce vectors that have considerable capacity to spare for spliced-in foreign DNA—packaging the entire spliced DNA molecule (λ DNA plus foreign DNA insert) into whole phage (Figure 7.4, overleaf). In fact, up to about 24 000 base pairs of foreign DNA can be thus accomodated.

Being able to clone relatively large fragments of DNA has another advantage. Say that we are constructing a complete gene library. Then the larger each separate fragment of foreign DNA, the fewer different fragments, and hence different clones, there will be. Thus the fewer the number of clones to be screened to find that containing our desired gene (Section 6.1). In fact, when constructing gene libraries, λ vectors, *not* plasmid ones, are often used. λ vectors are also used frequently in constructing cDNA libraries.

Needless to say, λ DNA can also be manipulated to produce expression vectors that can give high yields of foreign protein product when such phages are used to infect cells.

▷ Assume that a recombinant λ phage is to be used to infect *E. coli* and get the foreign protein product of a eukaryote gene carried by the phage. What sort of foreign DNA should probably have been used to construct the recombinant phage?

▶ As the foreign gene is eukaryotic it is probably a split gene. As here we require the protein product not the entire gene, a cDNA or a synthetic gene (i.e. minus introns) should have been spliced into the λ DNA in constructing the phage.

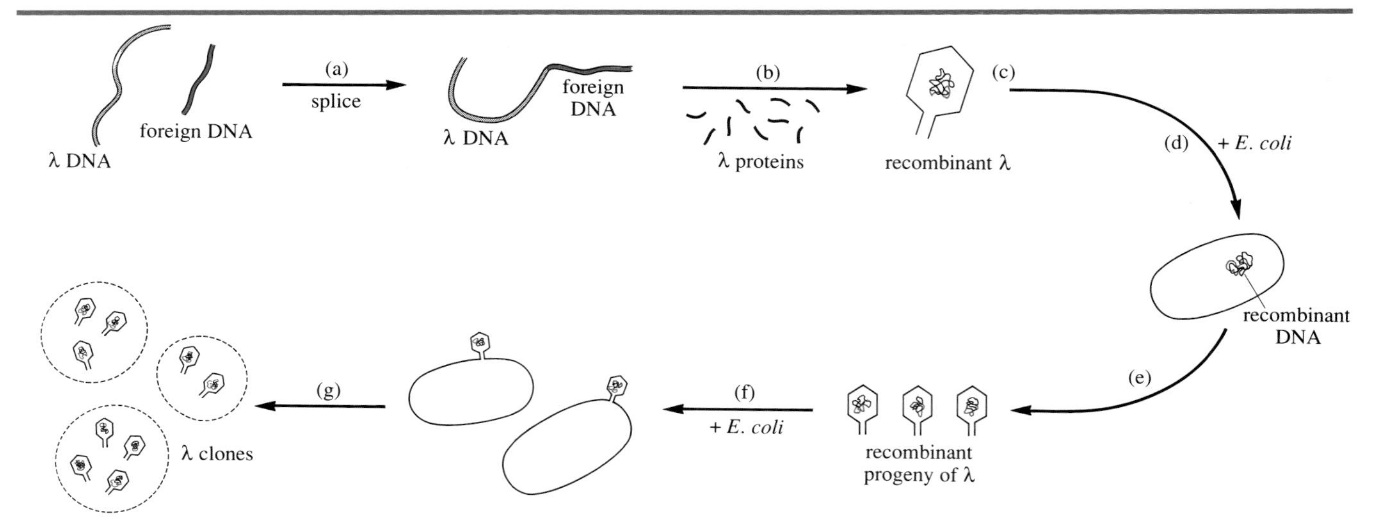

Figure 7.4 Using a λ vector to clone foreign DNA. (a) The foreign DNA (restriction enzyme fragment *or* cDNA *or* synthetic gene), shown in red, is spliced into DNA derived from λ. (b) The recombinant DNA is added to λ proteins *in vitro*, resulting in assembled, recombinant, phage (c). (d) The recombinant phage are used to infect *E. coli*, each phage infecting one cell. (e) The resulting progeny phage are also recombinant, i.e. contain recombinant DNA. (f) They can infect neighbouring cells, and so on, ultimately giving rise to clones of recombinant phage (g). The λ clones are screened to find the ones carrying the required gene. (As in Figure 7.2, the *E. coli* chromosome is not shown.)

7.2 Other host cells

Much of the early development of rDNA technology was done using *E. coli* as a host and it is still the organism where our repertoire of genetic engineering tricks is at its most sophisticated. Nevertheless *E. coli* cells are not the only hosts into which foreign genes can be transferred and their protein products produced. A number of other microbial cells, prokaryote and eukaryote, have been used as hosts, as have individual eukaryote cells derived from, *multicellular*, animals and plants. We will consider very briefly some such hosts. But given that *E. coli* represents such a well-developed system, we must first ask 'Why do we need other hosts anyway?'

7.2.1 Why other microbial hosts?

As we argued at the outset, one of the prime motivations for transferring foreign genes into microbes is our ability to grow microbes in large quantities and hence potentially to obtain readily the products of those foreign genes in large amounts (Section 3.4). As a particularly well-studied laboratory organism, *E. coli* was, and still is, an ideal experimental system in which to develop new techniques (Section 4.1). But when we move from laboratory to industry, other developments are necessary; broadly speaking, how to grow the microbes on an industrial scale. Some generalizations apply, but each organism is to some extent a rule unto itself and a good deal of trial and error is inevitably involved in scaling up the use of a microbe new to industry. Though it is the laboratory organism *par excellence*, traditionally *E. coli* has not been an important organism *industrially*. So the experience of growing it on a huge scale is lacking. Such experience is available for a number of other microbes, both prokaryote (like *E. coli*) and eukaryote.

▷ Name a eukaryote microbe that has long been used industrially.

▶ Yeast.

Among prokaryote (bacterial) microbes with an industrial track record is *Bacillus licheniformis*, grown in large amounts for its protease, an enzyme produced in hundreds of tonnes per year, worth tens of millions of pounds and used in washing powders and as a meat tenderizer (Table 3.2). Other bacteria grown in large batches industrially include *Corynebacterium glutamicum*, an important source of amino acids (Table 3.2 and Figure 3.3).

So experience on an industrial scale might be a good motivation for using such bacteria, or yeast, as alternative hosts to *E. coli* for foreign genes. There might well be other reasons for using microbes other than *E. coli*, some positive, some negative. We shall mention briefly just three.

1 Assuming it is the protein product itself of a foreign gene that we are after, not just a cell with altered metabolism, we want as high a yield of this protein as possible. Getting large amounts *per cell* is one aim to which developments such as finding powerful compatible promoters can contribute. But as well as producing the foreign protein, it also needs to be isolated from the cells and purified to some extent or other, free of host cell proteins. As mentioned above, sometimes high production per cell can complicate the isolation of native undenatured protein from such cells. If the protein were synthesized and then *secreted* by the cells into the growth medium, this might well simplify its recovery and purification. It so happens that some microbes such as members of the *Bacillus* genus and yeasts such as *Saccharomyces cerevisae* naturally secrete some of their own proteins. If they are used as hosts for foreign genes, with suitable guile on the part of the experimenter it can prove possible to get the foreign protein secreted too.

2 Among the major early protein products of genetically engineered microbes, ones with pharmaceutical use are likely to be prominent (Chapter 9); indeed some, such as human insulin, are already on the market. Food products, or substances used in food processing, are also likely to be important commercial outputs from such novel microbes. As substances either given medicinally or used in food or food preparation, the foreign proteins themselves need to be tested for safety; something that may need to be done anyhow for any new product with such intended use irrespective of its origins, whether natural or synthesized chemically or made via genetic engineering. It is also vital to ensure not just that the foreign protein is itself safe for use but that it is purified free of any other product of the microbe that might cause problems. It so happens that there is a powerful toxin in the cell wall of those strains of *E. coli* that are most useful for genetic engineering. So any protein product isolated from cells of such *E. coli* would need to be shown free of this component. This is a good (negative) reason for *not* using *E. coli* as a host for some products, as ensuring toxin-free material would necessarily increase production costs. It is also a good (positive) reason for using other microbial hosts such as, say, yeast, where a long history of industrial use to produce food products (beer) or as a food itself, has proved the host cell's inherent safety.

3 As you are aware, production of a polypeptide chain is not strictly speaking the final step in the path from gene to protein. For one thing, the polypeptide must fold up, perhaps in combination with other polypeptides, to give the final characteristic three-dimensional shape of the protein in question (Box 5.1). However, in many *eukaryote* proteins, but not all, the process is complicated by further events termed collectively **post-translational modification**. Some of these involve cleaving the polypeptide chain to give the final biologically active product.

▷ Can you recall one such eukaryote protein?

▶ Insulin. It is synthesized initially in the pancreas by translation of an mRNA to give a polypeptide called pre-pro-insulin. This undergoes cleavage producing pro-insulin. The final, biologically active molecule, insulin, is generated following further cleavage of the pro-insulin polypeptide.

Other post-translational modification involves systems of enzymes that can modify the R groups of some amino acid residues present in completed polypeptide chains. For example, short chains comprising linked sugars may be covalently linked to certain amino acid residues; other residues may be phosphorylated (i.e. have a phosphate group attached) or sulphated (i.e. have a sulphate group attached); yet other modifications may occur. These modifications do not occur to natural prokaryote proteins and hence no such post-translational modification enzymes exist in prokaryotes. *If* it proves important that a desired foreign protein has such modifications, then a prokaryote host, such as *E. coli*, will not give the desired fully-modified protein product. This may be a good reason for using a eukaryote microbial host, such as yeast, where *some* such modifications can occur.

However, it is important to note that eukaryotes seem to vary considerably in exactly what modifications they can perform. For example, though yeast can add sugar chains to R groups of amino acid residues of foreign eukaryote proteins it does *not* add the same sugars as occur in the natural product. And yeast cannot sulphate proteins at all. If a more precisely modified eukaryote protein is required it may be necessary to go to a host cell closer in evolutionary origin to the foreign gene—say a mammalian cell as host for a mammalian gene—but even here subtle differences in modification may exist. However, some proteins that in their natural state are post-translationally modified seem, fortunately, to be biologically active even without modification, or at least without some such changes.

7.2.2 Animal and plant cells as hosts

One of the most attractive features about rDNA technology is that it offers the opportunity to produce eukaryote proteins, that may be hard to obtain in bulk otherwise, in simple microbial systems (themselves prokaryote or, like yeast, eukaryote). All the advantages of microbes may then accrue—rapid growth, dealing with single-celled organisms rather than whole multicellular ones, and relative cheapness of production. Human insulin produced in bacteria is one such example.

It may thus seem ironic that on occasions we are led to clone animal genes in animal cells, perhaps to achieve efficient post-translational modification. Nevertheless, we are speaking of animal (or plant) cells *not* the whole animals (or plants). Essentially, what we try to do is to treat animal or plant cells much as we might treat microbes. This is made possible, in part, by a technique known as *cell culture* or **tissue culture**.

As the name implies, tissue culture, involves taking cells isolated from individual tissues of animals and plants and getting them to grow as single cells in laboratory glassware—human or rat liver cells growing and dividing in a dish, for example. For many years tissue culture has provided a way of studying certain features of animal and plant cells that would be difficult to study in the whole organism. It has also had some other practical applications for which a couple of examples will suffice.

- Animal cell tissue cultures are used for growing certain viruses on the way to producing vaccines.
- Shikonin, a bright red plant pigment from the roots of the shikon, *Lithospermum erythrorhizon*, is a substance used traditionally in Japan as both a dye and an anti-inflammatory agent (against burns and haemorrhoids). It can be also produced from tissue cultures of shikon cells. An early commercial outlet for this tissue-culture-derived shikonin was in a 'bio-lipstick' marketed by Kanebo, the Japanese cosmetics firm.

But in practice tissue culture is far from easy. Animal and plant cells are not naturally single-celled microbes and do not appear to take kindly to being plucked from their more communal habitat inside an animal or plant. Nevertheless, over the years tissue culture of a wide variety of animal and plant cells has been achieved. There is however, a price to pay, sometimes literally. Unlike most microbes, tissue culture cells often require complex media for their survival and growth, something that makes their cultivation expensive. They also tend to change in character, often differing from the native cells from which they were originally obtained (e.g. liver). However, though perhaps somewhat altered, stable *lines* (i.e. growing and dividing cultures) of such cells can be maintained. Compared to most microbes the rate of growth and division of tissue culture cells is paltry. Though it may be possible to grow tissue culture cells on the surface of nutrient jellies and hence select clones (colonies), this too is more difficult than the same technique applied to microbes (Figure 6.2). These and other factors combine to make tissue culture a tricky and expensive technique and one that industrially has yet to remotely approach microbes in scale—batches of over 1 000 litres would be considered large for tissue culture cells (as against microbial batches such as in Figure 3.3). Despite all these factors, tissue culture is a useful technique. Its usefulness has been expanded since the advent of rDNA technology. Tissue culture can provide the individual animal or plant host cells that may be needed for efficient production of proteins of certain cloned eukaryote genes. Let us take one example.

Tissue plasminogen activator (t-PA) is a thrombolytic, i.e. a blood clot dissolving agent. It does not dissolve clots itself, but when present in the bloodstream t-PA can activate a chain of events that leads to the dissolution of fibrin, a protein that forms a network that holds a blood clot intact. Thus t-PA has useful therapeutic value, such as in removing a clot (thrombus) that occurs, say, in coronary thrombosis, a major factor in heart attack. Native human t-PA comprises some 527 amino acids, contains sugar side-chains and has some disulphide bridges in its complex three-dimensional structure. cDNA for t-PA has been cloned in *E. coli* and expression to produce biologically active t-PA achieved.

▷ What does this tell us about the role of the sugar side-chains in t-PA?

▶ As sugar side-chains will not be added by *E. coli*, it tells us nothing about their role in t-PA, except in the negative sense that they do not seem absolutely vital for biological activity. They may have some more subtle role, such as how long t-PA remains active in the bloodstream, for example.

However, the yields of biologically active t-PA from *E. coli* have been disappointing, something that may be laid at the door of this prokaryote microbe's inability to achieve readily some post-translational events; in particular, disulphide bridge formation.

Fortunately, the gene (cDNA) for t-PA can be cloned and expressed in a number of mammalian tissue culture cells where reasonable yields of biologically active protein are produced. Such genetically engineered mammalian cells have provided the commercial source of t-PA, a product that has been approved for medical use since 1987. The inevitably high costs of using cultured mammalian cells, as against microbes, can be offset against the price that can be commanded by a useful pharmaceutical. As of the early 1990s, t-PA weighs in at some $2200 per dose (Section 9.3.2).

7.3 How to get the genes in and expressed

Having explained briefly some reasons why host cells other than *E. coli* may be needed, we must describe *very* briefly some techniques for introducing foreign genes into such cells and getting their products expressed there.

Once again, it is convenient to divide the process into three phases:

- getting the foreign gene
- introducing it into the host cell
- getting it expressed in its new surroundings to produce the foreign protein.

7.3.1 Getting the foreign gene

For this first aspect, the *host* cell is essentially irrelevant. In principle, the techniques available are the same for all foreign cells. The foreign gene (from the 'donor' cells) can be obtained as a fragment of DNA (Section 4.3) or as a cDNA (Section 5.3) or as a synthetic gene (Section 6.5). Of course, if either of the first two techniques are used we will still need, later on, a suitable method of identifying the right clone deriving from the engineered (*host*) cells, i.e. the clone containing the desired gene.

7.3.2 Getting the gene into the host cell

The second aspect, getting the gene into the host cell, is much more varied. To some extent, each different type of host cell requires a different technique. Fortunately, by comparing some of the variations with the methods I have described for *E. coli*, we can construct a brief summary.

As with *E. coli*, there are plasmids indigenous to some other microbes. These include certain species of bacteria (prokaryote) and fungi (eukaryote), including, notably, *S. cerevisae*. With suitable cutting and splicing, these plasmids can serve as vectors for transferring foreign DNA into such species, much as we use plasmid vectors in *E. coli*. Even where indigenous plasmids are not known in a particular species, it is sometimes possible to use plasmids from other species to construct usable vectors. For example, a plasmid from *Staphylococcus aureus* can be tailored to serve as a vector in *Bacillus subtilis*. There are no known plasmids truly indigenous to plants. But, as you will see shortly (Chapter 8), a remarkable 'natural genetic engineer' involving a *bacterial* plasmid can be used to great effect in plant cells.

Like λ in *E. coli*, viruses can be used as vectors in a variety of cells. This has been particularly important for gene transfer into animal cells. Perhaps the best known of such vectors are ones based on a virus called SV40. SV stands for 'simian virus' and indeed SV40 infects rhesus monkeys. In the laboratory SV40 is conveniently grown on kidney tissue culture cells from African green monkeys. Much like λ, tailoring SV40 DNA allows it to be used as a vector for foreign DNA and a number of SV40-derived vectors are known. Like λ, some SV40-derived vectors infect animal cells, replicate their DNA there and assemble with viral proteins to give progeny viruses; these progeny will carry any foreign DNA that was suitably spliced into the SV40 DNA and replicated thereby along with it (*somewhat* analogous to λ in Figure 7.4, though a different method is used to get the recombinant DNA packaged into virus particles to start with). This provides a method of cloning and replicating large amounts of a foreign gene—like λ, the gene is cloned in the virus not in the cells.

However, other vectors derived from SV40 behave somewhat differently. Instead of replicating their DNA inside cells and packaging this into virus progeny, they do not replicate. But they insert their DNA, *including any spliced-in foreign genes, into the host cell's own genome*, i.e. integrate it into one of the host's chromosomal DNA molecules. When such a host cell replicates its own genome the integrated virus-plus-foreign DNA *is* replicated as part of it and passed on to the daughter cells when the host cell subsequently divides. In this way, such SV40-derived vectors can be used to stably, permanently, introduce foreign genes into the genome of certain animal cell hosts. (This type of SV40-derived vector is not limited to monkey cells.)

▷ Why is the permanent, stable, nature of the integration important?

▶ If we wish to use animal cells as a source of a foreign gene and its protein product then ideally any such clone of genetically engineered cells that we obtain should be stable. That is, as the cells grow and divide in the culture they replicate their own DNA plus the integrated foreign DNA, so that all offspring cells continue to have and express the foreign gene.

Such integration of introduced DNA into the host cell genome is by no means unique to such SV40-derived vectors nor to animal cells. For example, in some instances, DNA transferred into yeast can integrate permanently into that cell's chromosomal DNA, be replicated along with it and thus passed on at cell division.

The mode of integration can also vary. In some instances integration seems to be at random sites throughout the host chromosomal DNA. In others, it is at specific sites where there seems to be some homology (i.e. similarity) of base sequence between introduced DNA (vector/foreign DNA) and the host chromosomal DNA.

▷ Does stability of a genetically engineered clone in *E. coli*, as produced using a recombinant plasmid, depend on integration of the foreign DNA into the *E. coli* chromosome?

▶ No, there is no integration needed. Stability depends on the fact that, though plasmids replicate independently of the *E. coli* chromosome, they are passed on to daughter cells at cell division. Thus each generation of cells will have plasmids and any foreign genes that were initially spliced into them (Section 4.2).

In fact some SV40-derived vectors can have a plasmid-like existence in monkey cells, replicating their DNA (plus any spliced in foreign genes) independently of the host cell's chromosomal DNA. However, such SV40-derived vectors are unstable, i.e. they are soon lost from the cells, and so such clones are unsuitable for long-term use for foreign gene products. However, vectors derived from another animal virus, bovine papillomavirus (which causes warts in cattle), can stably replicate with their spliced in foreign DNA and be passed on plasmid-like at cell division inside suitable host cells (bovine or rodent); no virus particles are produced and the cells are unharmed. This plasmid–*E. coli*-like situation provides an alternative means of achieving stability to that of chromosomal integration of vector-plus-foreign DNA.

There is one final point worth noting about this extremely compressed treatment of the wide variety of vectors needed to effect efficient transfer of genes into a wide variety of cells. As stated earlier, nowhere are our techniques as developed and as sophisticated as in *E. coli*, and that is why we have chosen to explain the *E. coli* system in some detail and to use it as our basis of comparison. But the ease of manipulation in *E. coli* has been exploited more directly—in producing vectors suitable for other host cells. A number of the vectors, plasmid or viral in origin, that are used in other cells, prokaryote or eukaryote, are in part constructed and manipulated in *E. coli*. Indeed some such vectors, so-called **shuttle vectors**, are hybrids containing DNA sequences that allow them to replicate in both *E. coli* and another host. So, for example, the manipulation to insert a foreign gene into a vector and subsequent DNA replication can be achieved readily in *E. coli*, while the gene expression to produce its protein product can be done in the other host.

The final form of vector that we shall consider is *no vector at all*. This relies on the fact that, given suitable conditions, many types of cell can take up DNA from the surrounding medium.

▷ Can you recall one such example?

▶ The uptake of plasmid DNA (natural or recombinant) by *E. coli* is itself such an instance. In this case, calcium ions provide the 'suitable conditions'.

So in essence, the host cells are mixed with the foreign DNA (say as restriction enzyme fragments or as cDNAs) which is *not* spliced into a vector. Given the appropriate conditions, the cells take up the DNA. And, *if* the absorbed DNA then happens to integrate into the host cell's own chromosomal DNA, as can occur in some cells, it will be replicated and passed to daughter cells along with the host cell's own genes. This vector-less technique does not work well with many types of cell. For one thing, the uptake may be inefficient. So might be the integration of the foreign DNA into the chromosomal DNA, something that, in the absence of a plasmid-like vector, is needed to achieve stable replication of the foreign DNA.

Tricks have been developed to increase uptake of the DNA. One technique, called *electroporation*, is applicable to a wide variety of cells and involves applying momentarily a high voltage to the cells. This electric shock seems to create temporary holes in the cell membrane and so allow DNA in the medium surrounding the cells a brief period of access.

7.3.3 Getting the gene expressed

Assuming that it is the gene product we are after, not just the gene, then having got the foreign gene established in its new host it remains to get the gene expressed efficiently.

I will not go into any detail here, but the principles are much as in *E. coli*. For example, having multiple copies of the foreign gene per cell should be a help, and there are ways of achieving this. Precisely how depends on the particular system (vector–host) in question. Likewise, vectors might be constructed that allow efficient expression of the foreign gene in its new host.

▷ Name something that this entails.

▶ Constructing a vector that has a promoter adjacent to the foreign gene that is recognized by the host cell's RNA polymerase and thereby allowing transcription of the gene (Section 5.1). As promoters are not universal, each time a new host cell system is developed suitable expression vectors may need to be found or new ones constructed.

Constructing a system that gives efficient gene expression can benefit from experience from other systems, notably *E. coli* again. But as with all new systems, ultimately only trial, and inevitably some error, can ensure success.

Activity 7.1 *You should spend up to 30 minutes on this activity.*

As will be evident by now, how precisely to clone a particular gene depends on the gene in question. This influences our choice of vector, whether to use the restriction enzyme technique or the cDNA one or a synthetic gene, which host cell to use, and so on. It also depends on whether we require just the gene itself or also its protein product and in what sorts of quantities. You now have an opportunity to consider some of these criteria, and others, for a particular gene. This activity thus tests knowledge of material in several chapters of this book.

Human Factor VIII (FVIII) is a protein that is involved in blood clotting. In the human body FVIII is synthesized initially as a single polypeptide 2351 amino acids long which then undergoes considerable post-translational modification to give active FVIII. Such evidence as there is suggests that the liver is the site of synthesis of FVIII. A hereditary absence of FVIII results in a type of haemophilia, haemophilia A. In recent years, sufferers from this haemophilia have been treated with FVIII prepared from donated blood plasma.

Say it is decided to try and produce human FVIII in a simpler system of single cells. To this end, those cells must be genetically engineered to contain a gene for human FVIII. The natural gene for FVIII is some 186 000 base pairs in length.

Answer the questions posed below about this proposed project and give brief written reasons for your answers.

(a) If the project was to attempt to produce FVIII in *E. coli*, should the actual gene derived as a restriction enzyme fragment be used or a cDNA?

(b) If a cDNA is to be used, what would be a suitable source of the mRNA?

(c) In fact, should *E. coli* be used as the host or another host? *If* another host suggest broadly what type of host.

(d) Suggest reason(s) why a source of FVIII other than from human blood plasma might be a good idea.

The answer to question (a) should be no more than 150 words in length; to question (b), around 50 words will suffice; for the others, no more than 100 words each (less for each if you use note form).

7.4 From single cells to whole multicellular organisms

So far, all of our treatment of the methods of rDNA technology have depended on inserting genes into *single cells*, whether those of naturally single-celled organisms (say, bacteria and yeast) or of single animal or plant cells in tissue culture. A prime motivation for doing this was laid out in Chapter 3—to create 'microbial factories'. This has been true in two senses.

In one, the foreign gene is transferred into a host cell not to 'improve' that cell *per se*, but to allow efficient production of the protein product of the foreign gene itself; it is this protein that we are after—human insulin, thaumatin and t-PA are examples referred to already.

In the other sense, the aim is somewhat different. It is to introduce a foreign gene so as to improve the output of a normal product of that *whole* cell—say, for example, the hypothetical transfer of starch-degrading enzymes into *S. cerevisae* to enable cheaper production of beer that is also lower in carbohydrate (Section 3.4). We said that one possibility might be to transfer specifically *several* genes coding for *different* enzymes involved in starch degradation into *S. cerevisae*. In fact, examples of this type might well present additional problems. Each of the different genes (each corresponding to a different desired enzyme) might well have to be specifically transferred separately, i.e. in a separate experiment. Cells from a clone with a newly acquired desired foreign gene would have to be subjected to a further experiment to transfer another, different, desired foreign gene into one or more of the cells; get a new clone with two desired foreign genes; and so on. So several 'rounds' of a protocol like that in Figure 4.2 might be needed; each 'round' would involve identifying just the right clone and so on. This could be very time-consuming and there is also the possibility that the host cells would be adversely affected by the demand of synthesizing and accommodating a number of different foreign proteins, each with its own effect on the host cell's metabolism. Indeed, in any gene transfer, of many or even a single foreign gene, there is always a degree of unpredictability as to what effect on the host cell's overall metabolism the new gene might have.

For either of the two objectives—foreign gene product itself required *or* changing the host cell's output of some normal product (e.g. alcohol by yeast)—the general methods are much the same, as both require efficient transfer of the foreign gene and its expression. Though in the first case, where the protein product itself is sought, the level of expression desired may be much higher. It is this type of case that we have mainly concentrated on. And even where the cells have not been true microbes, but have been of animal or plant origin, we have used them much as if they were microbes—have grown them as single cells in tissue culture, have attempted to use them as 'factories' for foreign gene products.

There is, however, a quite different type of genetic engineering of animal and plant cells—not the transfer of genes into single cells to be grown and cloned in culture, but the transfer of foreign genes to produce entire genetically engineered multicellular organisms. It is to this production of genetically engineered animals and plants that we now turn in Chapter 8.

Summary of Chapter 7

1 *E. coli* is not the only host for foreign genes, nor are plasmids the only vectors.

2 *Bacteriophage λ* has been much used as a vector for *E. coli*. It can tolerate *larger pieces of spliced-in foreign DNA*. It can be used to clone foreign genes in the phage itself—i.e. constructing gene or cDNA libraries in complete, packaged, phage.

3 *Other microbial hosts* may be needed because: they are better understood for *industrial use*; they present *no safety hazards* such as might the cell wall toxin of *E. coli*; they can *secrete certain proteins*; eukaryote microbes, such as yeast, may carry out some *post-translational modification of polypeptides*, such as *attaching sugars to amino acid R groups*.

4 Animal and plant cells can be grown as single cells in *tissue culture*. However the technique is tricky and relatively expensive as compared to growing bacteria or yeast.

5 Tissue culture cells can provide a source of animal and plant host cells. These may carry out *post-translational modification* needed to get some eukaryote protein products in *biologically active* form.

6 A variety of vectors have been developed for host cells other than *E. coli*, both for other prokaryote cells and eukaryote ones too. Some vectors are based on plasmids, others on viruses. *Some virus-derived vectors also behave like plasmids*, co-existing with but replicating independently of the host cell genome. Some cells can take up DNA without vectors.

7 In many host cells the introduced *foreign DNA ends up spliced permanently into the host cell's chromosomal DNA* where it can replicate and be passed on generation after generation of cell.

8 *Suitable promoters* can be found to get expression of foreign genes in a variety of host cells.

9 Whatever host cells vectors are to be used for, it is often convenient to carry out the manipulation of such vectors in *E. coli*. Some vectors, called *shuttle vectors*, can replicate in both *E. coli* and another host cell. These allow manipulation and production of the vector (plus foreign genes) to be done in *E. coli*, followed by transfer of the genes, via the vector, to the other host cell.

Question 7.1 Which of the following statements are true and which are false?

(a) λ can be used as a vector for large fragments of foreign DNA produced by restriction enzymes, but not as a vector for cDNAs.

(b) One advantage of using yeast as a host cell for foreign genes is that yeast can secrete proteins.

(c) One advantage of using yeast as a host cell is that it correctly attaches sugars to newly synthesized polypeptides.

(d) Some strains of *E. coli* are no good as hosts for the production of proteins that are to be used for human therapy because they contain a toxic substance.

(e) If we wish to clone an entire eukaryote gene that has introns and exons we can use either a restriction enzyme technique or a cDNA one.

Question 7.2 Considered as a host for the production of biologically active foreign *mammalian* proteins, which of the following is likely to be generally true of:
(a) *E. coli*, (b) yeast (*S. cerevisae*) and (c) mammalian tissue culture cells?

(i) Newly synthesized polypeptides can get sugars attached to them.

(ii) Restriction enzyme fragments of foreign DNA provide an adequate means of obtaining the necessary foreign gene.

8 Genetic engineering of plants and animals

In this chapter we go from genetic engineering of single cells to that of whole multicellular plants and animals. And, in many ways, the methods used for genetic engineering of whole plants and animals are similar to those used on single cells, whether bacteria, yeast or tissue culture (Chapters 4–7).

However, the multicellular nature of the intended hosts for foreign genes immediately introduces certain differences, as discussed below.

- The first, and most obvious, difference is that, unlike *E. coli*, the organism is not single-celled. So we need a method of ultimately introducing the foreign gene into the very many somatic cells in a complex organism such as an animal or plant. As for a single cell (which goes on to yield a clone; Figure 4.2h), the introduction must be stable, the foreign gene must not be lost. That is, as cells within the organism grow and divide, via **mitosis**, the foreign genes must be passed on to the daughter cells just like the normal host genome.
- The second difference concerns the reproduction of the organism as a whole. Here again stability is vital. That is, when reproduction occurs, we wish the foreign genes to be passed on, inherited by, the offspring. In some species of plants reproduction can be asexual—say, via runners as in the strawberry or tubers as in the potato or a 'gardener-induced' cutting of an African violet. But in many plants, and virtually all animals, *sexual* reproduction is the method of propagation. This means, that to be passed on at reproduction, any foreign genes must be present, following meiosis, in the germ cells (eggs and sperm), not just in the somatic cells. This is a complication absent in single-celled organisms where, in effect, the one cell is the sole 'somatic' *and* 'germ' cell.
- Finally, in multicellular organisms, though all cells carry the same genes, not all genes are expressed at all times nor in all cells; that is, their expression is *time-* and *tissue-specific* (Box 5.1). Assuming, as before, that we wish foreign genes to be expressed, it is likely that we would wish also that such differences between types of cell apply. For example, if we were transferring a foreign gene conferring blue pigment to a flower, we may not want blue stems too. Similarly, there would be little point in a plant expending energy in synthesizing in its roots large amounts of a foreign protein useful in photosynthesis.

With these points in mind, we now turn to the genetic engineering of plants and animals—first plants.

8.1 Plant regeneration

Most of the methods of genetic engineering of plants depend on the remarkable fact that many plants can be **regenerated** from single cells or small pieces of plant tissue (called **explants**). Given such cells or explants, complete mature sexually competent plants can be grown from them. In many instances regeneration proceeds via a *callus*.

In nature a **callus** is a clump of cells attached to each other, that can form on a plant wound. The cells in a callus are *undifferentiated* (i.e. in a sense, like embryonic cells). In the laboratory, calluses can be produced from explants of tissues that contain many undifferentiated cells, such as root tips or buds. Place these explants on a suitable solid medium (i.e. a suitable nutrient jelly) and the cells within the tissue divide to give a callus. A callus itself can be propagated merely by subdividing it. Change the conditions in the medium, notably the levels of certain **plant hormones** (also known as *plant growth regulators*), and roots or shoots emerge, leading ultimately to small plants or **plantlets** (Figure 8.1). Plant these out, allow them to grow and mature plants result.

Figure 8.1 Plant formation from a callus. Plantlets can be seen beginning to regenerate from this callus derived from a species of wild tomato. Note that several plantlets arise from a single callus.

However, if instead, a callus is put in a liquid medium and shaken, the clump breaks up. Single cells can be obtained and in certain media these will grow and divide.

▷ What would this give us?

▶ A plant tissue culture.

If plated out onto suitable solid media each single cell can further grow and divide to give a coherent mass of *attached* cells, a *callus*, not a colony of *individual* single cells as do microbes. Suitably treated, as above, such calluses will give plantlets.

▷ What will suitable treatment involve?

▶ The right levels of plant hormones.

Thus, in essence, as shown in Figure 8.2, we have an amplification cycle allowing mass propagation of plants—many new plants from one old one. Variations on this cycle exist.

Importantly, in principle, such techniques may allow us to produce *clones* of identical plants—many plants from a single callus (Figure 8.2f) which itself derived from a single cell (Figure 8.2d). (However, in practice, some genetic variation does arise among members of the same clone. How this happens is not known, as yet.) This cloning technique has already found commercial application. For example, oil palms with superior characteristics have been propagated by cloning. This *asexual* way of propagating a plant that normally undergoes sexual reproduction avoids the risk of losing any mix of superior characteristics during the 'sexual lottery' (Box 2.1).

So far, our discussion relates to regenerating plants from explants or single cells via calluses. It says nothing about genetic engineering of plants. But it stands to reason that if we can get foreign genes stably into callus cells, then we can regenerate whole plants bearing those genes. And, as implied in Section 7.3.2, it is possible to get foreign genes into plant cells—the question is *how*?

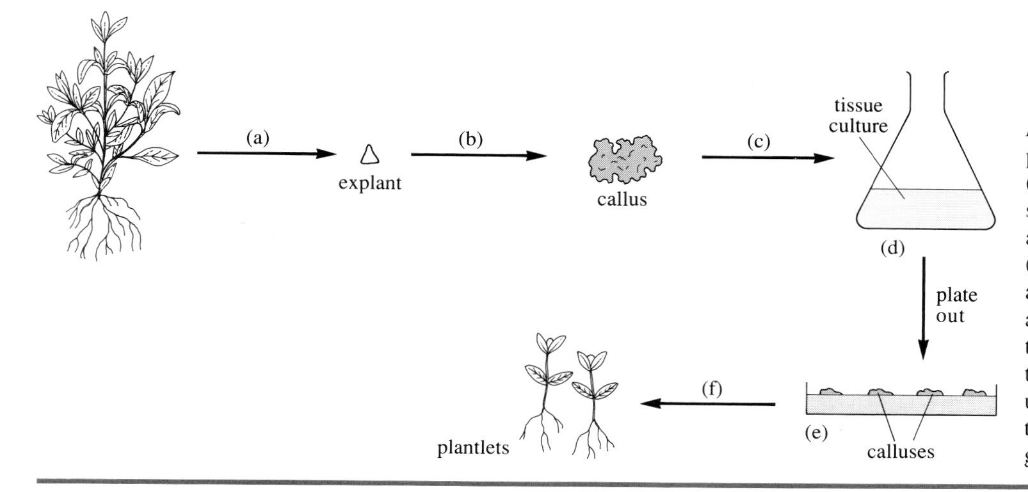

Figure 8.2 Mass propagation of plants via callus and tissue culture. (a) An explant is taken from a suitable plant tissue and (b) allowed to grow to form a callus. (c) When placed in liquid medium and shaken, the callus cells separate and divide to form a tissue culture (d). When plated out, each tissue culture cell gives an individual callus (e). (f) When taken and treated appropriately, each callus gives a number of plantlets.

Figure 8.3 Two varieties of crown gall disease.

8.2 The first genetic engineer?

Tobacco, potato, beans and most trees are examples of **dicots** (**dicotyledenous plants**); plants that have two **cotyledons**, two leaves in the seed. Many dicots are affected by **crown gall disease**. This manifests itself as a lump, a **gall**, at the site of a wound (Figure 8.3).

The cause is infection by a species of soil bacterium, ***Agrobacterium tumefaciens***. Galls are masses of undifferentiated plant cells. They are tumours and the calluses derived from them remain tumorous when cultivated *in vitro*, free of bacteria. They also synthesize substances called **opines**. These are not found in normal plant cells but are related chemically to certain unusual amino acids. Different strains of *A. tumefaciens* induce galls that produce different opines.

A. tumefaciens metabolizes opine (different strains metabolize their own particular opine) which it can thus use as its main nutrient and its sole source of energy. This metabolism depends on enzymes unique to *A. tumefaciens*, that are lacking in other soil microbes. Thus, having induced an opine-producing tumour, *A. tumefaciens* is free to enjoy its ill-gotten gains without competition from neighbouring species of soil-dwellers. *A. tumefaciens* achieves this cosy exploitation by genetically engineering the plant!

The *tumour-inducing* capacity of *A. tumefaciens* depends on possession of plasmids, known thereby as **T_i plasmids**. It is by transferring such plasmids into plant cells that *A. tumefaciens* induces gall formation. Following the transfer, *a specific portion of the plasmid DNA becomes integrated into the chromosomal DNA of the plant cell.* This portion of plasmid DNA is known as the **T-region** or **T-DNA**. It contains genes that code for proteins involved in gall formation and others that code for enzymes needed for opine synthesis. Other genes, such as ones needed for plasmid transfer and for opine utilization by the bacterium, are elsewhere on the plasmid.

Thus *A. tumefaciens* has long achieved what human genetic engineers are now so proud of—the transfer and expression of genes across a wide gap in the species barrier (Chapter 2), from bacterium to plant. Needless to say, now that we know this, we have latched on to its success.

8.3 Vectors from T_i plasmids

T_i plasmids are natural vectors transferring *bacterial* DNA (the T-DNA of *A. tumefaciens*) into plant cells where it becomes stably integrated into the plant cell chromosomal DNA, thereby replicating along with it. Adapting such natural vectors to carry in other foreign genes, of our choosing, is relatively obvious.

Question 8.1 Outline how a T_i plasmid *might* be used to carry a foreign gene into a plant cell and get that gene integrated stably into the plant chromosomal DNA. Ignore how the foreign gene itself is obtained.

A problem with merely using a T_i plasmid to carry a desired foreign gene into a plant is that the plasmid would presumably still induce tumours, galls, in the plant—not so desirable perhaps. So, in essence, what has been done is to remove, cut out, the tumour-inducing genes of T_i plasmids and use the resulting modified plasmid as a vector.

▷ How do you think such genes could be cut out?

▶ By using restriction enzymes and re-sealing the cut plasmid with ligase.

We can construct vectors which are hybrids between T_i minus its tumour-inducing genes and *E. coli*-based plasmid vectors. This allows most of the manipulations, including splicing foreign genes into the T-region, to be done in the simpler, better understood, milieu of *E. coli*. Essentially, we then take up the hybrid vector, with any desired foreign genes inserted (hence a *recombinant* plasmid), into *A. tumefaciens* and these are then used to infect suitable plant tissue. Infection results in the uptake of the recombinant vectors into plant cells and the subsequent integration of the T-DNA, complete with foreign genes, into the plant chromosomal DNA. It then remains to regenerate whole genetically engineered plants from the altered cells. Variations on the theme exist, but a simplified representation of what has become the basic and widely used procedure is shown in Figure 8.4.

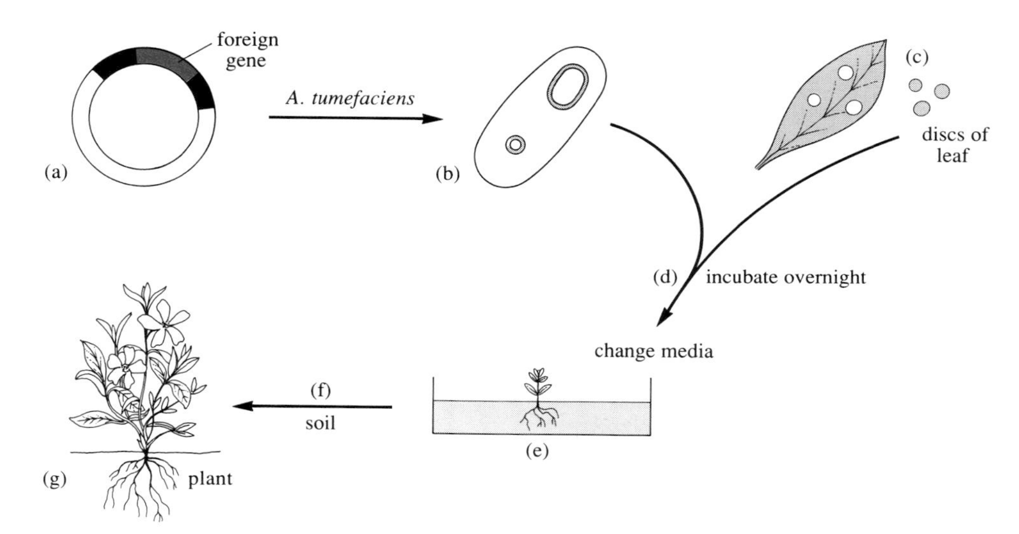

Figure 8.4 (a) The desired foreign gene (shown in red) is spliced *within* the T-region (shown in black) of a hybrid *E. coli* plasmid–T_i vector that lacks tumour-inducing genes; these manipulations are done in *E. coli*. This vector with foreign genes inserted is then taken up into *A. tumefaciens* (b). (c) Explants in the form of small discs are punched from a leaf and (d) incubated with these bacteria overnight. (e) After a number of changes of medium (notably to kill residual *A. tumefaciens* and alter levels of plant hormones) plantlets develop (via a callus). (f) These can be subsequently planted out in soil to give mature genetically engineered plants (g). The process, from infection (d) to planting out in soil (f), takes a month or two.

8.3.1 Getting foreign genes expressed in plants

As with other systems (Section 5.1), to get foreign genes *expressed* in plants, suitable promoters compatible with plant cells must be included alongside the foreign genes. Some such promoters are known and appropriate expression vectors can be constructed. But there is also the point unique to multicellular organisms about tissue-specific expression (see the beginning of this chapter), i.e. getting transferred genes expressed in some plant tissues and not in others.

It appears that in natural plants, in addition to promoters, there can be DNA sequences adjacent to structural genes that regulate (control) the transcription of those genes in particular tissues and under particular circumstances. Thus, for example, a gene coding for a protein that binds chlorophyll, the light-absorbing pigment of photosynthesis, is only fully active in chloroplast-containing tissues, such as leaves, where photosynthesis occurs. The gene activation is itself brought about by the stimulus of light and is dependent on *regulatory DNA sequences* close to the gene. Splicing of such **regulatory** (DNA) **sequences**, as well as a promoter, to foreign genes and transferring them into plants leads to the foreign genes being expressed only in leaves in the light.

We can, if desired, alter normal tissue locations. For example, we can attach *leaf-specific* regulatory sequences from tobacco to the gene for *patatin*, a major protein of potato *tubers*. Following gene transfer to give engineered potato plants, we can get the *transferred* patatin gene expressed in the *leaves* of those plants, not the tubers.

▷ What else does this experiment tell us about DNA sequences controlling tissue expression in plants?

▶ They are *not* totally **species-specific**, that is, not specific to one species alone—the tobacco leaf-specific regulatory sequences work in potato leaves.

Using stimulus- and/or tissue-specific DNA sequences, the expression of foreign genes can be manipulated at will, any gene in any tissue—in principle, that is, for we do not yet have complete knowledge or command of all the natural signals.

8.4 Cereal problems

You will have noted that we stated that crown gall disease occurs in many *dicots*. Yet many of the most important crop plants are not dicots but **monocots** (i.e. **monocotyledenous plants**, having a single cotyledon). Grasses (Graminae) are monocots and this includes cereals, three of which, wheat, maize and rice, top the list of the world's major crop plants.

In the laboratory, *A. tumefaciens* can be made to induce tumours on one or two species of monocot and T-DNA can be shown to be integrated into the genomes of some monocots, in general. But *A. tumefaciens* and its plasmids do not work well on monocots, cereals in particular, and so other ways of genetically engineering these major crop plants have been sought. Among them is the possibility of using vectors derived from plant viruses, some of which are known to infect monocots. It is early days yet in the development of plant virus vectors but, ironically, the best studied so far, cauliflower mosaic virus, infects dicots.

Another way of getting foreign genes into plant cells is *without vectors*, for example just allowing cells to take up foreign DNA, much as was described at the end of Section 7.3.2, a technique applicable to dicot *or* monocot plants. However, unlike animal cells, in plant cells the cell membrane is surrounded by a tough cellulose cell wall and this needs to be removed in order to allow the DNA access to the cell interior. Plant cells without their walls are called **protoplasts**. Access of DNA to protoplasts can be enhanced by techniques such as electroporation (Section 7.3.2).

If, following genetic engineering, protoplasts are placed on suitable solid media they can regenerate a new cell wall within about a week. These cells can then divide, form a callus and, ultimately, regenerate whole genetically engineered plants (Figure 8.2). But once again problems arise with cereals.

Protoplasts can be prepared from cereals and DNA can be got into them and thence into the plant chromosomal DNA. But regenerating plants from single cells (via protoplasts or not) has proven extremely difficult for cereals.

The regeneration of the agriculturally most important cereals—rice, maize and wheat—from single cells *has* been achieved. But it is probably fair to say that such regeneration is still a far from routine affair and is much more difficult than regeneration of, say, dicots such as tobacco or petunia or of certain other monocots such as asparagus. Nevertheless, however difficult, this regeneration, coupled with direct uptake of DNA, has recently allowed rice and maize to be genetically engineered.

In regenerating plants from single cells, each species probably needs its own variant of 'the right stuff'. Cereals have proven very recalcitrant, but recent successes suggest that with time and effort regenerating them from single cells (protoplasts), and hence genetically engineering them, *might* become eventually a routine event.

However, other techniques which introduce DNA and by-pass protoplast regeneration, some quite bizarre, might be used.

8.4.1 Biolistics

Take a preparation of foreign genes, place them in a cartridge, load a gun, take aim at the plant and fire—hey presto—foreign genes shot into plant cells! If you had proposed such an experiment, say, about ten years ago, it would probably have produced indulgent smiles, *at best*. But, in essence, that is what Klein and his colleagues tried in 1987—*and they succeeded*—thereby giving a whole new meaning to the term 'shotgun technique'.

In fact what they did was to coat minute tungsten particles of around 4 μm diameter with foreign DNA, load these particles into a special gun and use a gunpowder charge to fire them at high velocity into onion cells. Particles were found to have entered many cells and a foreign gene expressed there transiently. In principle, this combination of biology and ballistics—now sometimes dubbed (perhaps unfortunately) **biolistics**—could perhaps be used to get foreign genes into a variety of plants. In principle, it should work on dicots or monocots. However, it is not without problems itself. Nevertheless this and variants of the technique have been used.

One such variant, in which DNA-coated gold particles are 'shot' into cells using an electric discharge particle accelerator, has recently been employed successfully to introduce foreign genes into rice. The target (here, literally!) material was immature rice embryos obtained from mature plants. Following the bombardment, calluses could be grown and thence mature rice plants. Some such plants were found to be genetically engineered and capable of passing on the foreign genes to their offspring via normal sexual reproduction.

▷ What does this show about those plants?

▶ The introduced genes were present in germ cells.

And, most recently and most excitingly (June 1992, in fact), somewhat similar success has been reported for wheat. This, the most major of all crop plants, succumbed to the particle accelerator technique applied to callus derived from embryonic tissue. From such bombarded tissue, whole genetically engineered plants were ultimately regenerated. The gene transferred was one conferring resistance to a particular herbicide. The regenerated plants showed resistance, as did their offspring. Though successful, in common with other cereals, these are very early days for such genetic engineering of wheat. This 'demonstration' experiment is far from being an everyday procedure. And, as far as predicting when genetically engineered cereals with new commercially attractive features will be on the market, we remind you about Niels Bohr (Chapter 1).

8.5 Genetically engineered animals

In producing genetically engineered plants, the object was to get stable introduction of foreign genes. Genes that could be expressed *and* passed on to the next generation of *whole* plants via normal sexual reproduction. As we have seen, it is possible to introduce foreign genes into animal cells; the genes can be integrated stably into the host cell chromosomal DNA and the genes can be expressed (Section 7.3). The same is true for plant cells. But there is a crucial difference when it comes to whole genetically engineered organisms. While genetically engineered explants of plant tissue or individual plant cells can be used to regenerate whole genetically engineered plants, there is no such technique available for animals. Unlike some plant cells, individual animal cells cannot grow into whole animals. There is however one notable exception to this.

▷ What is this exception?

▶ The fertilized egg (zygote), the single cell from which each animal derives.

If we are trying to produce whole genetically engineered animals, then the zygote seems a reasonable place to start. Indeed, as far as genetic engineering of mammals (the animals that we will concentrate on) is concerned, this approach has proven the most successful so far. We can do no better than describe what has rapidly become a classic experiment.

8.5.1 Bigger — and better (?) — mice

In the 1980s Brinster and his colleagues in the USA carried out a series of experiments in which genes were injected into newly fertilized mice eggs. The most dramatic of these was undoubtedly the injection of a gene coding for rat growth hormone. It is the basic protocol for this experiment that we shall describe. We shall also use it to pose a number of questions about the logic of the protocol.

Growth hormones are proteins which, as the name implies, are stimuli for animal growth — too little can lead to dwarfism, too much to gigantism. Brinster and his colleagues isolated (i.e. cloned, see later) the gene coding for rat growth hormone. This was isolated as a gene not as a cDNA.

▷ What might this gene contain?

▶ As a eukaryote gene, in addition to coding sequences (exons), it might contain non-coding introns (Box 5.2). In fact the gene contains five introns.

The object of the experiment was not just to transfer the growth hormone gene into mice but also to get the gene expressed.

▷ Gene expression involves transcription and translation of the gene (Box 5.1). Name two things there must be to ensure that expression of this rat growth hormone gene occurs?

▶ 1 There must be an efficient promoter adjacent to the gene to allow transcription in the mouse cells (Section 5.1).

2 The sequences in the primary mRNA transcript corresponding to introns must be excised — i.e. post-transcriptional modification to give mature functional mRNA must occur (Box 5.2).

To ensure efficient transcription, the rat growth hormone gene was spliced to a promoter (I need not detail how this was done) known to be efficient in mouse cells. In fact this was a *mouse* promoter that is normally adjacent to the gene for mouse metallothionein-1, a protein that binds heavy metals such as cadmium and zinc. This promoter had been shown previously to allow transcription of another foreign gene (a virus-encoded one) when integrated into mouse DNA.

A preparation of this hybrid DNA (rat growth hormone gene plus mouse promoter) was then injected into newly fertilized mouse eggs. In fact it was injected into a *pronucleus*. Just after fertilization the egg contains two nuclei, prior to their fusion—a male one (from the sperm) and a female one (already in the unfertilized egg)—known as pronuclei. The male pronucleus is larger than the female one and therefore easier to inject into. The injection technique is shown in Figure 8.5.

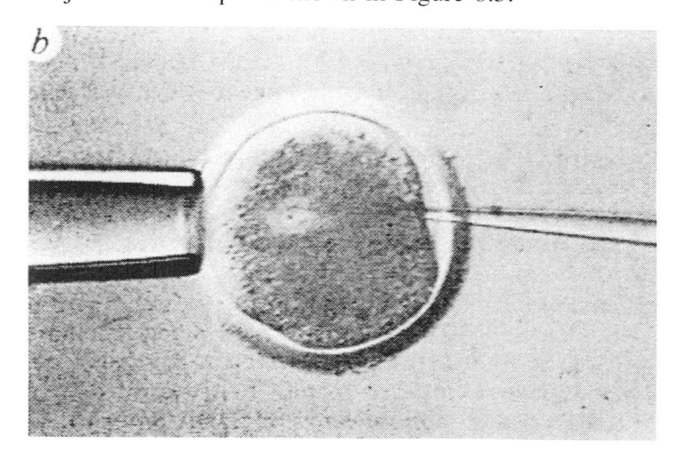

Figure 8.5 The fertilized egg is held by suction with a pipette (seen on the left) and the DNA is injected into a pronucleus of the fertilized mouse egg using a very fine needle (right).

▷ Why inject the DNA into a nucleus?

▶ Remember we wish ultimately to get stable introduction of foreign genes into a whole animal that will then pass these genes onto its offspring via sexual reproduction. This means that we must get these genes integrated into the chromosomal DNA and chromosomes are in the nucleus. Though DNA taken up generally into cells *may* end up in the chromosomes (this can happen for tissue culture cells; Section 7.3), injecting it directly into the nucleus at least gives it a flying start.

Each egg received about 600 copies of the promoter–growth hormone hybrid DNA. The injected eggs were then implanted in the uteri of surrogate-mother mice. Of the 170 treated eggs implanted, 21 came to full term. Seven of these could be shown to possess one or more copies of the rat growth hormone gene integrated into their chromosomal DNA. And, of these seven, six were significantly larger than littermates developed from untreated eggs. Mice with multiple copies of the rat growth hormone gene also had very high levels of growth hormone in the blood. Evidently, as hoped, transferred rat growth hormone genes were efficiently expressed in mice. And, when a similar experiment was done using *human* growth hormone gene attached to this promoter, broadly similar results were obtained (Figure 8.6). There was some evidence that tissue specificity was obeyed.

Question 8.2 What might these results suggest about the relationship between humans, rats and mice with respect to (a) post-transcriptional modification and (b) the biological effects of growth hormone? Assume that the human growth hormone gene is broadly similar to that from rat.

Question 8.3 Growth hormone is normally synthesized in the pituitary gland, metallothionein is synthesized mainly in the liver and kidney. When the mouse metallothionein promoter was spliced to the rat or human growth hormone gene it was attached to some other adjacent mouse metallothionein regulatory DNA sequences. Assuming that tissue-specific regulatory sequences operate much as do those in plants (Section 8.3.1), where would you expect the rat or human growth hormone to be made in the mice derived from the treated eggs?

However, one critical question remains. Is the transferred growth hormone gene passed on from parent to offspring? Some breeding experiments were performed, using engineered mice as one parent in each case. From these crosses, the answer appears to be yes, the gene can be passed on. And studies on the *pattern of inheritance* of the transferred rat growth hormone gene showed that it obeyed the classic laws of inheritance, those first elucidated by Gregor Mendel. This suggests that the gene was integrated stably into the mouse chromosomal DNA and was thereby inherited as any normal mouse gene.

Figure 8.6 The odd couple. ‘Supermouse’ and her sister are from the same litter and around 24 weeks old. There is about a twofold difference in weight between the mouse engineered with *human* growth hormone gene (that on the right) and her normal sister; dramatic enough to make the cover of *Science*, a major American journal. What, in some senses, is the ‘genetic explanation’ for this difference in weight, arches graphically about them. (From *Science*, 1983, **222**.)

▷ What else do such breeding experiments show about the introduction of the growth hormone gene at the fertilized egg stage?

▶ The gene ends up in both the somatic cells *and* the germ cells. Its *expression* to produce larger mice from treated eggs is evidence for its presence in somatic cells (Figure 8.6). The gene being passed on to offspring of these mice, via normal sexual reproduction, is evidence of its presence in their germ cells.

The technique pioneered by Brinster and his colleagues is not in theory the only possible one for getting genetically engineered animals. However, in practice, it has in many ways proven the most successful so far and has been used for transferring a number of different genes into a number of different species of animal. As you will see in Sections 10.6–10.8, these include farm animals such as sheep. We now also have a quite impressive array of DNA sequences that control adjacent gene expression in a variety of specific tissues. Splicing one or other of these adjacent to our foreign gene allows us a good deal of control over where, i.e. in which tissues, that gene will be expressed. Brinster's technique thus seems to have fulfilled our three criteria unique to multicellular organisms detailed at the beginning of this chapter: getting the foreign gene into all the somatic cells, via mitosis; getting the gene into germ cells, following meiosis, and passing the gene on at sexual reproduction; expressing the gene specifically in some tissues and not others.

However, before ending this topic, there is one final point we should make—how do we get the right foreign gene? Think about the problem for a moment.

In bacteria, we generally did not get the right gene *in advance*. We could do shotgun experiments—transfer in an array of genes as fragments of DNA or as a mixture of cDNAs, *then* screen the resulting clones of engineered bacteria (Chapter 6). But this is hardly on for animals.

DNA injection into a (fertilized) egg is tricky; each injection must be done individually; each egg must be implanted into a surrogate mother; and each must be brought to term. Not a technique that lends itself to producing millions of randomly engineered animals followed by a search for the one bearing the desired gene.

The answer is that we clone the desired gene *in advance*. We use a simpler system, such as tissue culture cells or bacteria, to clone an array of DNA fragments or cDNAs. We screen these to get the clone with the desired gene, and use this clone as our source of that gene which we can then manipulate further (e.g. add a suitable promoter or tissue-specific control DNA) *prior* to introducing it into the fertilized egg. All that is then required is to examine the animals produced to see that the gene has in fact been transferred successfully—not to see which gene it is. Hence, in the example we gave, we referred to injecting a specific gene, that for growth hormone, not a mixture of DNA fragments or cDNAs. Significantly, once again *E. coli* can often be gainfully employed. Many of the manipulations to knock the foreign gene plus attached control sequences etc. into shape can be done in this familiar system before we inject it into eggs. In fact an *E. coli* plasmid was employed in just such manipulations—having served this purpose much of the DNA was removed, and just the DNA with the desired growth hormone gene plus regulatory sequences etc. was injected.

Similar arguments can also be applied to plants. Who wants to grow and screen millions of plants? Though, where plants are to be regenerated from single cells, it may be possible to use mixed DNA (fragments or cDNAs) and screen the engineered cells to find the ones with the right gene, then grow plants from just those cells.

8.6 Transgenic plants and animals — where now?

As we have seen, it is possible to produce whole genetically engineered plants and animals — what are nowadays usually referred to as **transgenic** plants and animals. Yet such techniques are still relatively new, dating from the early 1980s. And problems still remain.

For example, the techniques referred to involve integration of foreign genes into the animal or plant chromosomal DNA. This indeed can be achieved, but integration is random, essentially anywhere within the host DNA. This can lead to two problems.

- First, different integration sites might lead to different efficiencies of gene expression — indeed differences in growth hormone level are seen between different tissues and different animals in Brinster's experiments and may well be related to particular sites of integration.
- Secondly, the integration may disrupt the functioning of an important host cell gene, an undesirable and possibly lethal side-effect.

So we may need ways of directing genes to specific integration sites, something that can occur naturally in some organisms such as yeast. Some such methods have been developed but others are needed.

Doubtless our ability to engineer animals and plants will increase rapidly, at least if the remarkable advances made in the last decade are anything to go by. Problems such as integration sites, and others, will yield to further developments in our skills for manipulating genes. But when dealing with whole complex multicellular organisms, knowing the genetic tricks is unlikely to be the whole answer. When applied to pigs, the growth hormone protocol produced arthritic animals — the gene transfer was successful but the effects on the animals' overall physiology could not be foreseen (hence the reservation in the title of Section 8.5.1) — a physiological *and* ethical problem (see Section 8.7). For new developments, trial and error, and hence uncertainty in prediction (Chapter 1), will never vanish entirely from such science.

8.7 From laboratory to market-place

We have now largely completed our examination of some of the basic techniques of genetic engineering begun in Chapter 4. This allows us to evaluate more critically the practical 'everyday' applications of genetic engineering with which we are chiefly concerned, as stated at the outset of the book. In particular we shall look at applications to medicine in the next chapter and to plant and animal breeding in Chapter 10.

But no scientific technique is sole arbiter of its own application. However powerful the technique and however desirable the application, the path from research laboratory to hospital, farm, supermarket or wherever is often a tortuous one. Many stimuli and stumbling blocks can serve to speed up or slow down the practical applications of genetic engineering. So, before embarking on our analysis of such applications, we ask you to consider briefly what kinds of things in general such boosts and barriers to innovation might be. To do this, have a go at the following activity.

Activity 8.1 *You should spend up to 30 minutes on this activity.*

Part 1

The first part of this activity concerns Extract 8.1, which you should read, and then answer questions (a)–(d) below.

(a) What was the motivation(s) behind developing the OncoMouse?

(b) Why was patenting important to DuPont?

(c) On what general grounds were the quoted objections to the patent based?

(d) What do you think that the general motivation/agenda of the protesting groups might be?

Your answers should be in note form, supported by evidence from the extract and around 30 words each. Use broad terms—ethical, commercial, etc.

Part II

Extract 8.1 illustrates *some* of the sorts of factors that may help or hinder a number of developments in genetic engineering. However, in other developments there may well be other factors. For the second part of this activity we suggest that you list other possible factors that might affect certain developments. Again summarize each factor in about 30 words. An additional clue is that many such factors may not be unique to genetic engineering; they may affect the development of any new product or process, to some extent or other.

Extract 8.1 From *The Times*, 15 October 1991.

Patent granted for cancer mouse

By Nigel Hawkes
Science Editor

THE European Patent Office has ruled that life-forms can be patented. The landmark ruling, allowing the granting of a patent to a genetically engineered mouse originally developed at Harvard University, has been issued by the patent office's board of appeals. Animal welfare and environmental groups criticized the decision yesterday.

The creature is known as the OncoMouse because its genes were modified by inserting oncogenes which cause cancer in humans. Because of its extra genes, the mouse is predisposed to develop cancer, making it a useful tool for researchers studying the disease and testing anti-cancer drugs.

The OncoMouse, marketed by the DuPont chemical company, sells for $100 compared with just over $1 for a regular laboratory mouse. It can be relied on to develop tumours, and then die, within about three months. Similar mice have been developed as models for studying heart disease, AIDS and genetically inherited diseases.

The OncoMouse was patented in the United States in 1988, but the process in Europe has been lengthy. The ruling of the board of appeals could open the doors to a flood of patent applications for other life forms, plant and animal.

The procedure of the patent office allows the presentation of a formal legal opposition to the granting of the patent. Compassion in World Farming and the British Union for the Abolition of Vivisection said yesterday that they would do this, in co-operation with other groups in Europe.

The groups said granting patents on genetically engineered animals 'displays a total disregard for animal life and fails to see that each creature is a sentient being, capable of suffering. That this patent should be granted on a mouse designed to suffer is an appalling example of scientific myopia'.

The groups said they believed a patent application had been filed on a chicken genetically engineered so that it would grow faster, and the males would produce sperm earlier, making the breed commercially profitable sooner.

Companies involved in biotechnology say that they need the protection of patents if they are to justify the research needed to produce improved crop plants and more productive animals. If the results of their work are unpatentable, they argue, there will be little point in doing it, since others will be able to copy it without any payment.

As even the single development described in Extract 8.1 shows, the interplay between factors affecting the application of genetic engineering can be quite complex. Consider the *motivation* for producing the OncoMouse.

From one standpoint, the project is based on *humanitarian* grounds—i.e. research aimed at curing cancer. Yet the groups opposing DuPont would consider *their* viewpoint the humane one; presumably considering that of DuPont, the scientists who developed the mouse and the patent office as 'an appalling example of scientific myopia'. Thus we can see how ethical issues can enter into such developments. Is it right ethically to inflict suffering on mice in order to potentially alleviate the suffering of humans? My point is not to pronounce on the rights and wrongs of such arguments but merely to indicate how the *very existence* of such ethical questions can affect the uptake of certain aspects of genetic engineering. Of course such questions are not unique to applications of genetic engineering—any use of animal experimentation can elicit similar questions to some degree or other, and a similar range of responses. However, there are broader ethical concerns which are perhaps more peculiar to genetic engineering—issues about the 'designing of organisms', 'tinkering with nature' and the like. These broader ethical questions will crop up again elsewhere.

The very concept of patenting an animal, exercising 'ownership' over a particular type of living organism, is also something that may cause ethical qualms. Yet from a commercial standpoint, patenting can be seen as justified, indeed necessary, to encourage such developments, as well as less ethically contentious ones too. Which of course brings us to the other motivation for the development of the OncoMouse—*money*.

Much of the basic science underlying medical advance is first done in non-commercial, publicly-funded research laboratories in universities, government research institutes and the like. Yet the later stages of **R & D (research and development)** needed to turn such science into usable medical devices, notably pharmaceuticals, and their marketing, is done more often by commercial organizations, part of what nowadays is sometimes called the *health-care industry*. As commercial organizations, such companies seek to make a profit, or, at the very least, cover the costs of R & D and marketing, something that can be very substantial, as we shall see (Section 9.1). DuPont might therefore seek to profit from the OncoMouse and hence their desire for patent protection. But a closer reading of Extract 8.1 might suggest a wider commercial reason than the OncoMouse itself.

As the opening paragraph makes clear, this was a 'landmark ruling', so DuPont's objectives (and those of the actual patent holders, Leder and Stewart of Harvard University) quite likely included helping establish the *principle of patentability* of genetically engineered animals. As the last paragraph of Extract 8.1 states, this is something that is seen as necessary by companies engaged in such developments. By definition almost, landmark cases are worth fighting for; something presumably perceived by the opposing groups too. So their opposition also was part of an agenda wider than OncoMouse alone (Activity 8.1d). Lose this one and 'the floodgates might open' to other similar exploitation of animals. Thus in a wider context, the outcome of this case can be seen as *stimulating* analogous developments (giving them a green patent light)—though whether the case will in fact establish precedent remains to be seen. On the other hand a win for the opponents would have erected a *barrier* to such developments: less certainty of patents—risk of failing to gain proprietary rights—less attractive commercially.

So this brief extract has already revealed some of the richness, the complexity, of the factors affecting developments in genetic engineering—various motivations/stimuli (humanitarian and commercial), arguments about patent law and ethical issues. Of course other developments will have their own particular mixes of factors affecting them; some factors in common with the OncoMouse project and other, different, ones. You have already considered other possible factors in Part II of Activity 8.1. But analysing the important factors in particular projects, as in Part I of Activity 8.1, also illustrates something else—the usefulness of examining particular examples, actual developments, as a way of teasing out the stimuli and stumbling blocks. Analysis like that in Activity 8.1 can become an important aspect of helping learn how to sort the wheat from the chaff, to distinguish feasible projects from those less so, something that in Chapter 1 we saw as a central aim of this book. However, real, convenient examples do not always exist, evidently true where 'predicting the future'—a tricky task (remember Niels Bohr)—is concerned. But, where necessary, we can *invent* suitable examples, so-called **scenarios**, and analyse them. *Scenario analysis* is sometimes favoured by political analysts, military strategists, economists, lawyers and business analysts, among others. Like them, we can use this 'What if such and such happens...?' game as a way of analysing the main issues and attempting to predict outcomes where the application of genetic engineering to medicine and plant and animal breeding is concerned.

Summary of Chapter 8

1 Producing whole genetically engineered multicellular animals and plants has some technical problems different from engineering single cells: *all the cells* in the organism must receive the foreign gene, it must be passed on from cell to cell at *mitosis*; the gene must be *passed on* to other generations of organisms at *sexual reproduction*; the gene must be *expressed in some tissues* and not others.

2 Genetic engineering of plants depends often on the ability to *regenerate* whole plants from *single cells* or tissue *explants*.

3 Plant regeneration often proceeds via an *undifferentiated mass of cells* called a *callus*.

4 *Plantlets* can be produced from a callus by altering *hormone levels* appropriately.

5 *Agrobacterium tumefaciens*, the cause of *crown gall disease*, is the source of important vectors.

6 The tumour (gall)-inducing capacity of *Agrobacterium tumefaciens* is by virtue of a plasmid called T_i *plasmid*. A region of the plasmid DNA, the *T-DNA* or *T-region* is inserted into the plant cell chromosomal DNA during natural infection.

7 *Vectors* that have *lost their tumour-inducing capacity* can be derived from the T_i plasmid. Foreign genes inserted in the T-DNA of such vectors can be carried into plant cells and then integrate into the host cell DNA.

8 Whole plants can be grown from cells or explants treated with *A. tumefaciens* carrying *recombinant plasmids* bearing foreign DNA.

9 Foreign genes can be expressed in plant cells by providing suitable attached *promoters*. *Tissue-specific expression* can be achieved by also including tissue-specific *regulatory (DNA) sequences* adjacent to the structural gene.

10 A combination of regeneration techniques and *A. tumefaciens*-mediated gene transfer can work well in *dicots* (e.g. tobacco, potato, etc.) but not generally in *monocots* (e.g. cereals).

11 *Protoplasts*, plant cells stripped of their cell walls, can take up *DNA*, and this is true for monocots as well as dicots. But some *monocots*, *cereals* in particular, are *hard to regenerate* from protoplasts; some can be regenerated, though (e.g. rice).

12 *Biolistics*, the shooting of DNA carried on tiny particles, is another way of getting foreign DNA into plants.

13 Whole *animals cannot be regenerated* from tissue explants nor from single cells.

14 *Injecting DNA* into a *pronucleus* of a *newly fertilized egg* is a way of introducing foreign genes into animals. The animal derived from such an egg contains the foreign gene in its chromosomal DNA. Given a suitable promoter and regulatory DNA

sequences, the foreign genes can be *expressed in a tissue-specific manner.* They can also be *passed on to offspring by sexual reproduction.*

15 Such techniques were pioneered in experiments transferring genes into *mice*, notably genes for *rat or human growth hormone*. The expression of such genes was manifest dramatically by *greater growth* of animals derived from treated eggs.

16 Many manipulations of foreign genes, including their *cloning* to get *one specific gene*, can be executed in simpler systems (e.g. bacteria) prior to injection into eggs (or transfer into plants).

17 Further work is needed on techniques to produce genetically engineered (i.e. *transgenic)* plants and animals. For example, ways of getting *integration* of foreign genes into *specific chromosomal DNA* sites are desirable.

18 A number of factors, connected only loosely to the science itself, may affect the *application* of genetic engineering to 'everyday' needs. These include, among others, *commercial* and *ethical* considerations. Such *factors* may serve to *stimulate* or *hinder* such applications.

Question 8.4 Which of the following statements are true and which are false?

(a) Monocots are easier to regenerate from tissue explants than are dicots.

(b) The male pronucleus is easier than the female pronucleus to inject DNA into because it is larger.

(c) Opines are substances produced by *A. tumefaciens* and not by normal plant cells.

(d) Integration of foreign DNA into host cell chromosomal DNA is generally at specific chromosomal sites in transgenic animals and plants.

Question 8.5 Assume that protein U is synthesized exclusively in unicorn liver and that a unicorn has recently been captured. The gene for this protein is transferred to a donkey by injection into a fertilized donkey egg followed by implantation of the resultant embryo into a surrogate donkey mother. Before the injection, the gene for protein U is spliced to the regulatory DNA sequences isolated from adjacent to a donkey structural gene for a protein known as donkein. In the resulting transgenic donkey, protein U is found exclusively in the brain.

(a) What is the likely explanation of these results?

(b) What does it tell us about donkein?

Question 8.6 Consider two hypothetical projects: (i) genetically engineering *E. coli* to produce the protein product of a human gene and (ii) genetically engineering wheat with a bacterial gene to confer insect resistance. Considering *only* the current state of techniques discussed in Chapters 4–8, which project would be likely to be achieved first and why?

9 Medical applications of genetic engineering

In Chapters 9 and 10 we move away from the specific techniques of genetic engineering to the *applications* of these techniques. We shall be considering two very broad areas: medicine and agriculture. Clearly, these are very large sectors, and we cannot hope to cover all aspects of them in great detail. We have therefore opted for a selective approach, dealing with a few topics in detail, while at the same time trying to put them in a broader context of the industry as a whole. In Chapter 10 we will be looking at some applications of genetic engineering in agriculture, but for now let us consider the ways in which genetic engineering can provide answers to problems in the medical sector.

9.1 The pharmaceutical industry

In the short term, the impact of genetic engineering on medicine is likely to be greater than anywhere else. This is largely for two reasons, humanitarian and commercial. The whole *raison d'être* of the pharmaceutical and medical 'business' is to prevent and cure diseases. Genetic engineering, as we shall see, offers new approaches to this, and indeed has already provided some new solutions to disease-related problems. But it should never be forgotten that the pharmaceutical industry is just that: an industry, which must make profits, or fail.

The pharmaceutical industry worldwide is a $multi-billion commercial enterprise. Worldwide sales in 1990 were estimated at $150–170 billion. Antibiotics alone account for more than $8 billion worth of sales annually. The industry has been dominated over the past 40 years by the 'big five' nations: the USA, Britain, France, Germany and Switzerland. These countries lead the field in terms of the development and introduction of new pharmaceuticals, with Japan very much the up-and-coming nation. It is widely believed that the success of pharmaceuticals in these countries is due to both a favourable academic climate, with a good foundation of basic research, and a favourable political and economic climate. The reason why the latter is important is that developing and marketing a new drug is a very lengthy and expensive process (Table 9.1)—the total cost may be in excess of $200 million.

Table 9.1 Development of a drug from discovery to market. (Source: Office of Health Economics, 1983)

Step	Probability of success/%	Proportion of total cost/%
chemical synthesis	—	15.8
screening	0.01	13.5
initial production	3	0.3
preclinical trials	13	57.6
initial clinical use	12	0.2
controlled clinical testing	25	5.6
confirmatory clinical trials	50	4.5
regulatory affairs	90	2.0
licensing authority	95	—
product launch and reviews	—	0.5

You can see from Table 9.1 that the most costly part of the process is the testing (for safety and effectiveness) before the drug is ever used clinically. It is very important to carry out this testing. All drugs can be dangerous; sometimes using a drug is a balance between 'killing' the disease and killing the patient. Only if the benefits of the drug outweigh the drawbacks (such as side-effects) will the drug be licensed for use. A new drug can be the product of more than ten years of basic research, and from the time of patenting a new substance to its general availability can take another ten years. Small wonder, then, that drugs are expensive! The pharmaceutical industry worldwide spends $billions each year on research and development ($15 billion in 1988), and this money must be recovered in drug sales. Not all drugs are going to be best-sellers, like tissue plasminogen activator (t-PA), a drug that helps to dissolve blood clots (see Sections 7.2.2 and 9.3.2), and which had sales of more than $200 million in 1990; so drug pricing has to be realistic, which generally means high.

The pharmaceutical industry is very proud of its record of innovation, and the 6000 or so approved drugs on the market worldwide demonstrate the industry's ability to develop and sell its products. Many drug companies have long-standing associations with the organic chemical industry. Such companies tend to undertake massive series of chemical syntheses of hundreds of chemical derivatives of a basic compound known to have some biological activity (Table 9.1). The many different derivatives of penicillin (one of the first antibiotics) are the results of a largely chemical synthetic strategy. Many of the derivatives produced in this way will be biologically useless, so a screening procedure must be undertaken to pick out the substances with biological activity. Once a promising substance has been detected, enough must be made for it to be tested—generally about 500 kg. In organic chemical terms, this is not a lot of substance, and it could be produced by a small-scale laboratory process. So pharmaceutical companies have their roots firmly embedded in *laboratory-scale* production, and are therefore well equipped to take on board genetic engineering techniques, which, initially at any rate, are all carried out on a very small scale. Yet these small volume, genetically engineered, products can be extremely valuable. So they are really an ideal target for pharmaceutical companies, and, indeed, such companies were among the earliest to exploit genetic engineering, with good results, as you will see shortly.

The new techniques of genetic engineering, outlined in earlier chapters, when applied to the foundation of basic research upon which the pharmaceutical companies depend, also allow us to learn more and more about the *molecular basis* of disease, with an increased likelihood, if not of a cure, at least of a safe, effective treatment. Genetic engineering also raises the possibility of effectively curing **genetic diseases**—that is, diseases which arise from faults in the genetic material—by altering the defective genes themselves, or by substituting correctly functioning copies of them. The possibilities are so exciting that it is little wonder that science fiction writers and journalists have exploited this rich seam to its limits. We hope to give you a realistic picture of what is actually possible now, and what is likely to become possible over the next few years. We will begin with a look at the two basic arms of the pharmaceutical industry: *diagnosis*; that is, the identification of a disease, and *therapy*, the treatment of it. We start with diagnosis.

9.2 Diagnosis

There are two stages in the medical process. First, the patient's reported signs and symptoms must be translated into a recognized disease; that is, a diagnosis must be made. Second, once the diagnosis has been made, treatment can begin, with the use of drugs if this is warranted. It is important that diagnosis be as accurate as possible so that the most appropriate form of treatment may be given. Diagnosis can be a very difficult area: a broken limb is quite easily diagnosed by both doctor and patient, but complaints such as lassitude and headaches can indicate a multitude of different conditions. In cases such as these, the patient may be subjected to a battery of tests, some of them biochemical, collectively termed **diagnostics**. In the pharmaceutical industry, diagnostics include not just tests for diseases, but also tests for various factors, such as blood type, where there is a range of normal phenotypes between which it is vital to differentiate. The victim of a traffic accident, for example, may need a blood transfusion, but if the wrong blood type is given the patient will die.

More accurate and sensitive diagnostics are constantly being sought. In general, the more sensitive a diagnostic test is, the earlier the disease can be identified, and the better will be the chance of a cure. Some diseases are, like broken legs, quite easy to diagnose because they show characteristic signs. Sugar diabetes, for example, is one of these.

▷ What is the characteristic sign of sugar diabetes?

▶ Sugar excreted in the urine.

This sort of characteristic is called a **molecular marker** for the disease. A high level of sugar, or more precisely glucose, in the *urine* is a marker, but elsewhere it might not be.

▷ Where and when is it normal to have high levels of glucose?

▶ In the blood shortly after a meal.

Thus it is important to indicate the source of the sample taken for clinical analysis.

A great deal of effort is directed towards finding molecular markers for diseases. Rather than the patient being admitted to hospital for a series of lengthy, sometimes invasive, tests, a small sample of blood, say, could be taken and subjected to a whole battery of biochemical analyses. The identification of (theoretically) as little as one molecule of a disease marker would be enough to clinch the diagnosis. In diseases such as cancers, where the rapid spread means that early, accurate diagnosis is vital, much research is directed towards finding suitable molecular markers. Progress is slow: for breast cancer, the most common form of cancer among Western women, preliminary diagnosis is still made by the woman herself feeling a lump—hardly a 'high-tech' method!

The ideal diagnostic, then, is one which is reliable, accurate, easy to use (eliminating the need for expensively equipped hospital laboratories), having a reasonable shelf-life, and widely available to the at-risk population. If self-diagnosis is to be undertaken, this means that the diagnostic must be available over the counter.

▷ Can you think of one diagnostic that fits this description?

▶ Modern pregnancy testing kits.

The marker molecule for pregnancy is a hormone called human chorionic gonadotropin (hCG). It is made by the placenta, and does not occur in non-pregnant women. Like glucose in diabetics, excess hCG is passed out through the urine, where it can be identified. Thus a pregnancy test involves a woman testing her own urine for the presence of hCG, which, if present, gives rise to some kind of easily recognized colour reaction.

The advances that genetic engineering could bring to diagnostics rely on the accuracy of specific marker molecule recognition. Two types of molecular recognition can be exploited here: recognition of specific *proteins,* and recognition of specific *genes.* The detectors used to pick these out are *monoclonal antibodies* and *gene probes* respectively. We discuss these below, beginning with monoclonal antibodies.

9.2.1 Diagnosis using monoclonal antibodies

Antibodies are part of the body's highly sophisticated defence mechanism, the **immune system**. They are proteins formed by specific cells in response to a foreign molecule, usually another protein, that has entered the body by chance. Collectively, foreign molecules that elicit the production of antibodies are called **antigens.** Antigens are often proteins that do not occur within the healthy body, but which *do* occur when the body is 'challenged' by a disease. For example, a protein antigen might be part of the cell wall of an invading bacterium.

▷ Do you think an antibody would be produced against glucose in diabetes?

▶ No. Glucose is a normal body constituent, and would not be recognized as foreign.

Antibody-forming cells can 'remember' antigens to which they have previously been exposed, and, if the antigen is presented again, they can mount a rapid attack to that specific antigen. Antibodies act by binding specifically to their target antigen, thereby inactivating it and causing a molecular aggregate to form which can be digested by other cells of the immune system. The aggregates are composed of many antibody–antigen complexes. Antibodies are Y-shaped protein molecules which consist of several subunits (see Figure 9.1). They are produced in warm-blooded animals (birds and mammals) by specialized cells—in mammals these cells are found in the spleen. One such cell can produce 2000 antibody molecules per second. The high specificity of antibodies is the result of expression of *specific* DNA in these antibody-producing cells. A competent immune system can produce antibodies against *any and every* foreign molecule that the animal encounters in its lifetime—this could amount to many millions of antibodies, each one different from all the rest. How is this diversity brought about?

The most obvious answer is that an antibody-producing cell contains a huge number of different genes, each of which codes for one specific type of antibody molecule. When a particular antigen is encountered, the appropriate gene is switched on, and the specific antibody is produced. Unfortunately, there are a number of overwhelming

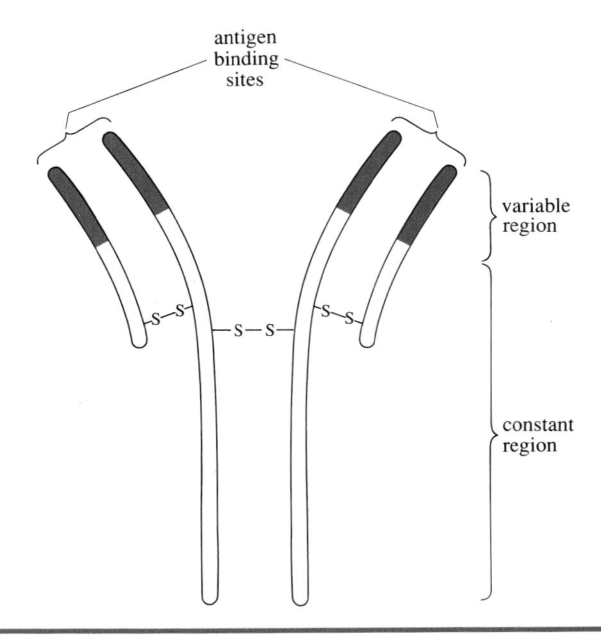

Figure 9.1 Basic structure of an antibody molecule. The subunits are joined via disulphide bridges to give the characteristic Y-shaped molecule with an antigen-binding site at the end of each 'prong'. Notice that the antigen-binding sites are formed from the *variable* regions of the antibody molecule. Other parts of the molecule consist of relatively *constant* amino acid sequences.

objections to this solution. First, there are only around 100000 structural genes. (Remember that much of the genome appears to have no informational function—so-called 'junk' DNA; Box 4.1.) But even if *all* of these genes coded for antibodies (and obviously, they do not), there would still be an order of magnitude too few genes to account for all the possible antibodies that an organism can produce. Second, animals are able to produce antibodies against synthetic molecules (such as complex organic chemicals) which have never existed in nature. How could genes exist which code for antibodies against such compounds? Deeper thought, then, suggests that the obvious answer is not the correct one.

Antibodies all share a common basic structure: it is only the antigen-binding site (Figure 9.1) which shows such amazing variability. So perhaps the answer lies in the way in which antibody diversity is generated. (Immunologists make some good jokes about the Generator Of Diversity, or GOD.) Some progress has been made over the past 15 years in understanding how a relatively small number of genes can give rise to such a vast range of antibodies (see Figure 9.2). It turns out that the DNA sequences coding for the various regions of antibody molecules, *domains*, can be shuffled—rearranged within the chromosome—to give many different combinations, resulting in a large number of possible outcomes from a limited set of building blocks. What is not yet known is how the cell 'knows' precisely *which* of the possible combinations will give an antibody which can bind specifically to a new antigen. Some information must pass from the antigen to the DNA of the antibody-producing cell regarding the structure to be recognized, but the nature of this information, and the way in which it is transmitted, is still shrouded in mystery. Transcription and translation of the rearranged DNA produces different polypeptide chains (Box 5.1) which will fold to give different shapes at the antigen-binding site, the 'business end' of the antibody molecule. It is this shape difference that guarantees the specificity of antibody–antigen reactions. Specific DNA rearrangements occur in 'ancestor' spleen cells in response to the challenging antigen. All the descendants of a particular ancestor spleen cell (a particular clone of spleen cells) will have the *same* antibody DNA rearrangements, and so make only *one* specific antibody. But a blood sample from a warm-blooded animal contains the output of millions of different clones of spleen cells, and so will contain millions of different antibodies, corresponding to all the antigens to which that animal has ever been exposed.

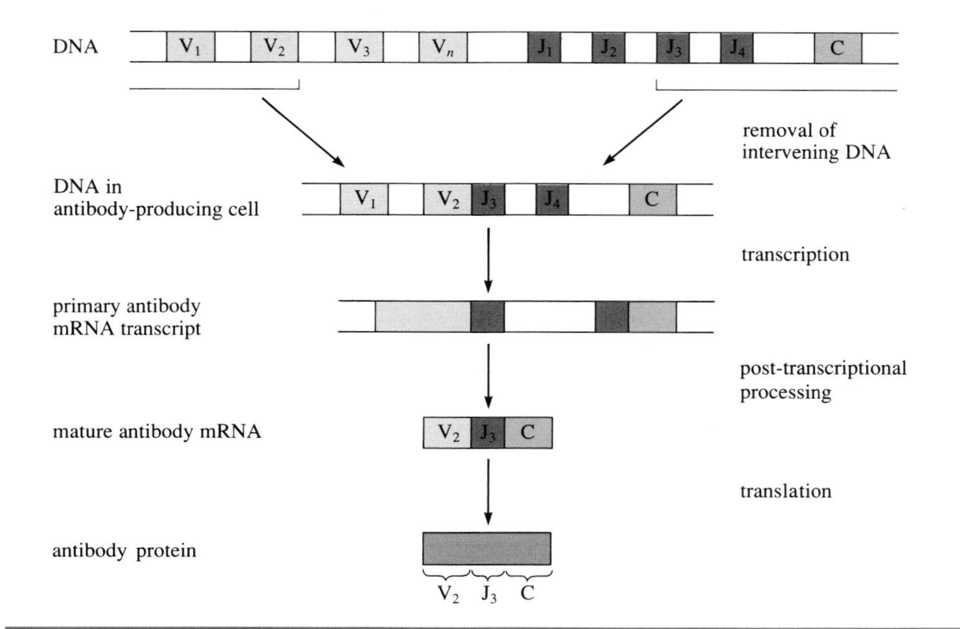

Figure 9.2 DNA rearrangements within spleen cells that are the ancestors of antibody-producing cells. The first step is the removal of the sequences separating the required V (variable) and J (joiner) regions (in this case, from the end of V_2 to the start of J_3). This is followed by transcription of the antibody DNA, producing the primary antibody mRNA transcript. The intron and, in this case, one of the two remaining J regions are then removed (post-transcriptional processing) to give the mature antibody mRNA, with the required V, J and C regions now joined together. Translation of this mRNA gives the corresponding antibody protein.

The use of antibodies for *therapy is* a well established procedure, and we shall discuss it further in Section 9.3.2. Traditionally, horses were used to produce antibodies against a specific disease-causing agent. This was because horses are large, and contain large volumes of blood from which a rough purification of antibodies could easily be achieved. The horse would be 'challenged' with a particular antigen (by injecting a sample of the antigen), resulting in the production of large quantities of the corresponding specific antibody. After a few days, blood which by then had a relatively high concentration of the desired antibody, was then taken from the horse. The horse blood, as we suggested above, comprised a mixture of antibodies against *all* the antigens the horse had ever encountered, and the only way to purify a *specific* type was to use the challenging antigen itself to 'pull out' those that reacted specifically with it. However, in the 1970s, a technique was developed in Cambridge which allowed production of large quantities of *pure* antibody of a chosen specific type. The ability to produce these so-called **monoclonal antibodies** depended on improvements in tissue culture techniques. The method is outlined in Figure 9.3. Although the technique works well, and can be done in species other than the mouse, its major drawbacks are that it is difficult and slow.

Nevertheless, monoclonal antibodies have made a great impact on the *diagnostic* market — reagents based on them are available to diagnose various cancers of colon, lung,

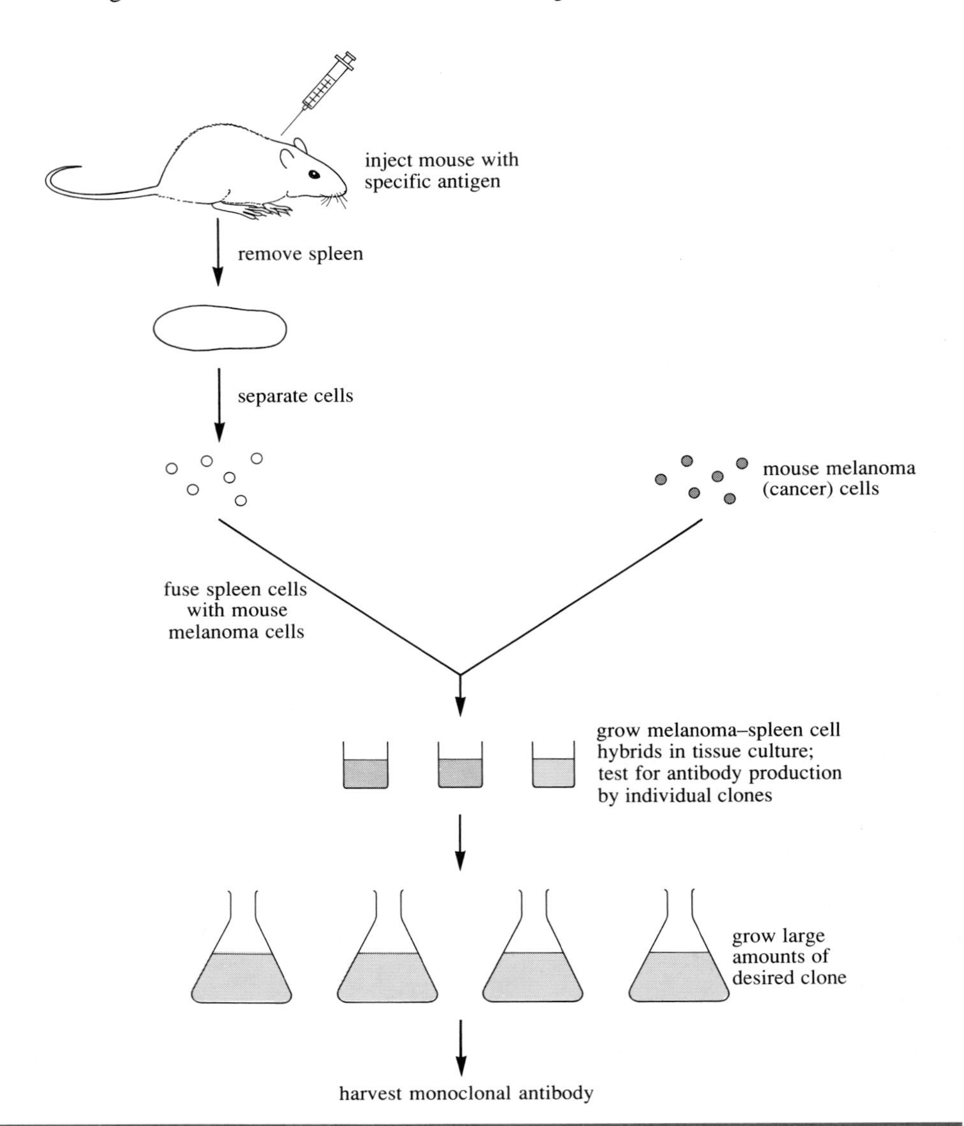

Figure 9.3 Main steps in the production of monoclonal antibodies. The antibody-producing mouse spleen cells do not survive in culture, but the mouse cancer cells can grow and divide indefinitely. The hybrid cells resulting from the fusion of these two cell types have the desirable properties of both of them: they secrete antibody with the same antigen specificity as the parent spleen cell and they proliferate in culture if properly cared for. The hybrids are cultured and screened to identify which clone (or clones) is producing the desired antibody (shown here in pink). The desired clones are then grown on a larger scale and the antibody subsequently harvested from the culture medium. The entire process takes several weeks.

breast, ovary and prostate, and other diseases. The reagents utilize the recognition by the antibody of specific protein markers for these diseases. Clinical samples are exposed to the diagnostic monoclonal antibodies, which have been 'tagged', for example with a fluorescent dye. If the molecular marker protein (the antigen) is present in the sample, the antibody will recognize and bind to it. After washing the sample, the antibody–antigen complex can be detected by the presence of the fluorescent dye.

▷ Why is it necessary to wash the sample?

▶ Washing removes the fluorescently labelled antibody which has *not* reacted with the target antigenic marker molecule. If the sample were not washed, it would be impossible to tell whether the monoclonal antibody had bound or not, and hence whether the diagnostic marker molecule were present or not.

The diagnostic tests for all the diseases mentioned above consist of taking clinical samples from the patient, such as blood, urine, saliva, etc. and testing these *in vitro* with the reagent. However, there are some diagnostic tests which involve introducing the reagent *into the patient's body*—an *in vivo* test. This might be done, for example, to detect by imaging techniques whether any secondary cancerous tumours had arisen from a primary one. If an antibody which specifically recognized a protein on tumour cells were injected into a patient, it would attach to tumour cells no matter where they were. If the antibody had been labelled with a fluorescent or radioactive tag, a body scan would identify the position(s) of the antibodies, and hence of the tumours. In this case it would obviously be impractical to try to include a washing step; fortunately, the blood circulation automatically removes any unbound antibody.

▷ Can you foresee a problem in injecting antibodies made in, say, a rabbit, into a human?

▶ The human immune system would recognize the rabbit protein as foreign, and therefore make its own antibodies to *it*, thereby disabling the reagent!

Thus, if it is necessary to test a patient *in vivo* with an antibody, ideally that antibody should itself be a human protein, so that it will be able to carry out its diagnostic function without hindrance. Although in theory it might be possible to make monoclonal antibodies from live human spleens, at present it is not feasible to do so. So how can monoclonal antibody technology be applied to human diagnosis?

As we have explained (Figure 9.3), a particular monoclonal antibody is produced by only *one* spleen cell and its descendants, i.e. a clone of spleen cells. Because the specificity of antibodies resides in the precise DNA rearrangements that take place in the ancestors of antibody-producing cells, it is clear that the machinery in these cells is able to recognize what are presumably very complex signals related to which bits of DNA to move, which to transcribe, and which to translate. You will therefore be surprised to learn that recently it *has* become possible to produce specific *monoclonal-like* antibodies in bacteria.

The part of the antibody molecule responsible for the specific binding, the variable region (Figure 9.1), can actually bind *in the absence* of the rest of the antibody molecule (the constant region). In this case it is known as a **single-domain antibody**, or, colloquially, a **mini-antibody**. Mini-antibodies do not bind as strongly or as specifically as complete antibodies, but they work well enough for most applications. The *rearranged* gene coding for the variable region, a single domain, can be removed from specific spleen cells (usually from a mouse), and inserted into *E. coli*, where it will be transcribed and translated to give a properly folded and functional mini-antibody.

▷ Why is it important to get a properly folded molecule?

▶ The binding specificity depends on the exact molecular shape of the antigen binding site (i.e. the mini-antibody). Without the correct shape, which depends on how the protein chain is folded, the mini-antibody will not be able to bind to the specific antigen.

Thus effective mini-antibodies can be produced in a couple of days instead of the weeks required to isolate a conventional monoclonal antibody (Figure 9.3). Moreover, yields can be increased to around 0.5 grams of antibody protein per litre of culture fluid, and the antibodies produced in this way do not themselves appear to be very antigenic.

▷ What does this mean?

▶ The mini-antibodies are non-human. However, even though they are foreign, they do not cause such a strong immune response in humans as do most intact, foreign substances.

With an increasing amount of knowledge of antibody structure, including amino acid sequence data, the leading group in the field, at the Medical Research Council's Molecular Biology Laboratory in Cambridge, have been able to go further and develop a range of 'humanized' mini-antibodies. Here, certain mouse amino acid sequences, not *directly* involved in antigen binding, have been chemically replaced by sequences exactly the same as those found in the antigen binding sites of normal *human* antibodies. These are expected to be much better tolerated by patients, because they appear less foreign to the human immune system. Humanized mini-antibodies look certain to make a major impact for *in vivo* diagnostic use, although as yet no products are available.

9.2.2 Diagnosis using gene probes

As we stated above, the advantage that genetic engineering can bring to diagnostics relies on the accuracy of marker molecule recognition.

We have already discussed the recognition of protein molecules; now we will discuss the recognition of specific genes.

▷ How might it be possible to recognize a specific gene?

▶ By the phenotype which it generates, in particular by the presence of its protein product; or, more fundamentally, by the gene's *specific base sequence*.

▷ How could a base sequence be recognized?

▶ By virtue of complementary base pairing. The use of a gene probe to detect a gene (Section 6.4.2) is a supreme example of specific molecular recognition.

By using gene probes it is possible to detect whether an individual is carrying a particular gene. This has applications in three broad areas: diagnosis of the presence of a foreign gene, for example from a bacterium or virus; diagnosis of a gene associated with an inherited disease; and diagnosis of a mutated gene characteristic of cancer. We will look at these applications in turn.

Detection of a foreign gene

Microbes, being living organisms, contain their own hereditary material which is very different in base sequence from, say, human DNA. Although there are some sequences which have been conserved through evolution, these are few and far between when, for example, bacteria and vertebrates are compared. It might therefore be imagined that there exist DNA sequences which are found *only* in, say, a disease-causing microbe. The presence of such a characteristic sequence in a clinical specimen would be a clear indication that the disease-causing organism (the *pathogen)* was present in the patient from whom the sample was obtained. That is, the foreign DNA sequence would be a molecular marker of the disease.

One traditional way to diagnose a suspected infectious disease is to take a clinical sample and attempt to culture *in vitro* any microbes present within it. This can be a lengthy and difficult business, as many pathogens are reluctant to grow outside the body. Indeed, it is common clinical practice not to bother *identifying* the pathogenic microbe unless it cannot be eradicated with the speedy application of a broad-spectrum antibiotic.

Many diseases not treatable by antibiotics can be treated by specific drugs, but in these cases it *is* important to know the identity of the pathogen, so that the appropriate drug can be administered. For these diseases, then, genetic engineering techniques can offer a rapid, and highly specific diagnosis: the identification of pathogen-specific DNA sequences in clinical samples.

Gene probes can be used not just to identify pathogens in clinical samples, but also as a screening method to detect the presence of *latent* pathogens, i.e. those which can reside inside a host's body for some time without causing disease symptoms. This sort of pathogen is usually a virus, which can exist for many years inside the host's cells, thereby escaping the notice of the immune system. Only when the virus starts to proliferate, and escapes from the cells, will the immune system perceive the foreign antigens, and by this time it may be too late to prevent the disease.

▷ Can you think of a well-known example of a latent viral infection?

▶ HIV (human immunodeficiency virus). Many people carry the latent virus for years; only when the virus becomes activated will they succumb to AIDS.

A gene probe test for HIV is currently available, and many people from the 'at-risk' population—which includes all sexually active people—are electing to be screened for the virus. In the future, it may become an essential part of every individual's health program, much as vaccination has now become (see below).

Detection of an inherited disease

It has been known for many years—indeed, centuries—that some diseases seem to run in families. You will be able to think of many examples of such diseases. Some are so mild that we may not even consider them as diseases, such as colour blindness. Others are much more severe, such as Duchenne muscular dystrophy or Huntington's disease, for example. With a few exceptions (for example, see Figure 9.4), only since the 1950s have serious studies been made of the familial incidence of such diseases. We can deduce from such pedigrees whether the defect is caused by one or more genes, and whether a disease-causing allele is dominant or recessive to the wild-type, normal allele (see Box 9.1).

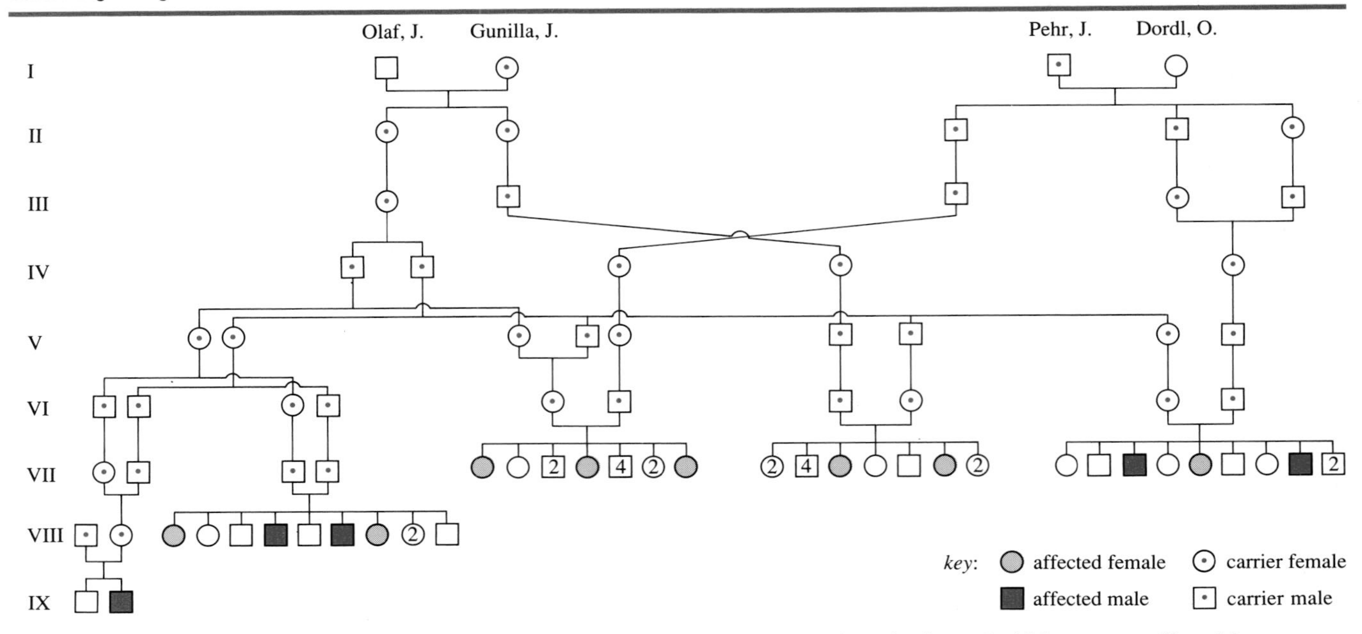

Figure 9.4 Pedigree of four Swedish families, derived from two original couples born in the early 18th century, affected by a recessive mutation which, in the homozygous state, leads to mental defect, spastic limbs and scaly red skin. (Roman numerals down the left-hand side denote the generations. Numbers inside the circles and squares denote numbers of offspring.)

9.1

Box 9.1 Basic genetics

Most animals and plants, including humans, are genetically diploid, that is, they carry *two* sets of chromosomes. One set is inherited from the individual's mother, and the other from its father. Each set carries a copy of all the genes necessary for that individual; this means that we have *two copies* of each of our genes. Although the two copies are very similar, they are not necessarily identical. As a result of past mutations, the exact base sequences *may* be slightly different in the two copies. Different forms of genes are known as *alleles*. It is combinations of different alleles that make us all individuals: we all have eyes, coloured hair, enzymes for digesting fats, etc., but we may all be slightly different in these respects. Some of the differences that can arise are so extreme that the gene does not produce a functional protein. An example of this is albinism, where a gene coding for an enzyme involved in the synthesis of melanin (a black pigment) is defective. The resulting individual has no skin pigmentation. Albinism is a condition that is not immediately life-threatening, but the vast majority of mutations result in alleles so defective that the individual carrying them cannot survive. This forms the basis of natural selection, and as the result of many thousands of years of evolution most of the individuals alive today have two of only a few 'permitted' alleles for each gene.

For reasons beyond the scope of this book, the presence of some alleles of a gene in a cell masks the activity of others. Alleles which mask others are called **dominant alleles**; those which are masked are called **recessive alleles.** Thus, in such cases, only the phenotype of the dominant allele will be evident. If a cell contains two copies of the dominant allele, it is said to be **homozygous** dominant. If it contains two copies of the recessive allele, it is homozygous recessive. This is the only time when the recessive phenotype will be evident, as there is no dominant allele to mask the recessive ones. When a cell contains one copy of the dominant allele and one copy of the recessive allele, it is said to be **heterozygous.** But because the recessive allele has been masked, the cell will be indistinguishable from a homozygous dominant one: that is, its appearance, or phenotype, will be that dictated by the dominant allele.

As mentioned above, each individual inherits one set of genes, arranged on chromosomes, from its mother, and one from its father. But each individual of a species has a particular number of chromosomes that it *must* have: too many or too few chromosomes result in abnormality or death. In humans the chromosome number is 46, made up of 23 pairs of chromosomes. One member of each pair

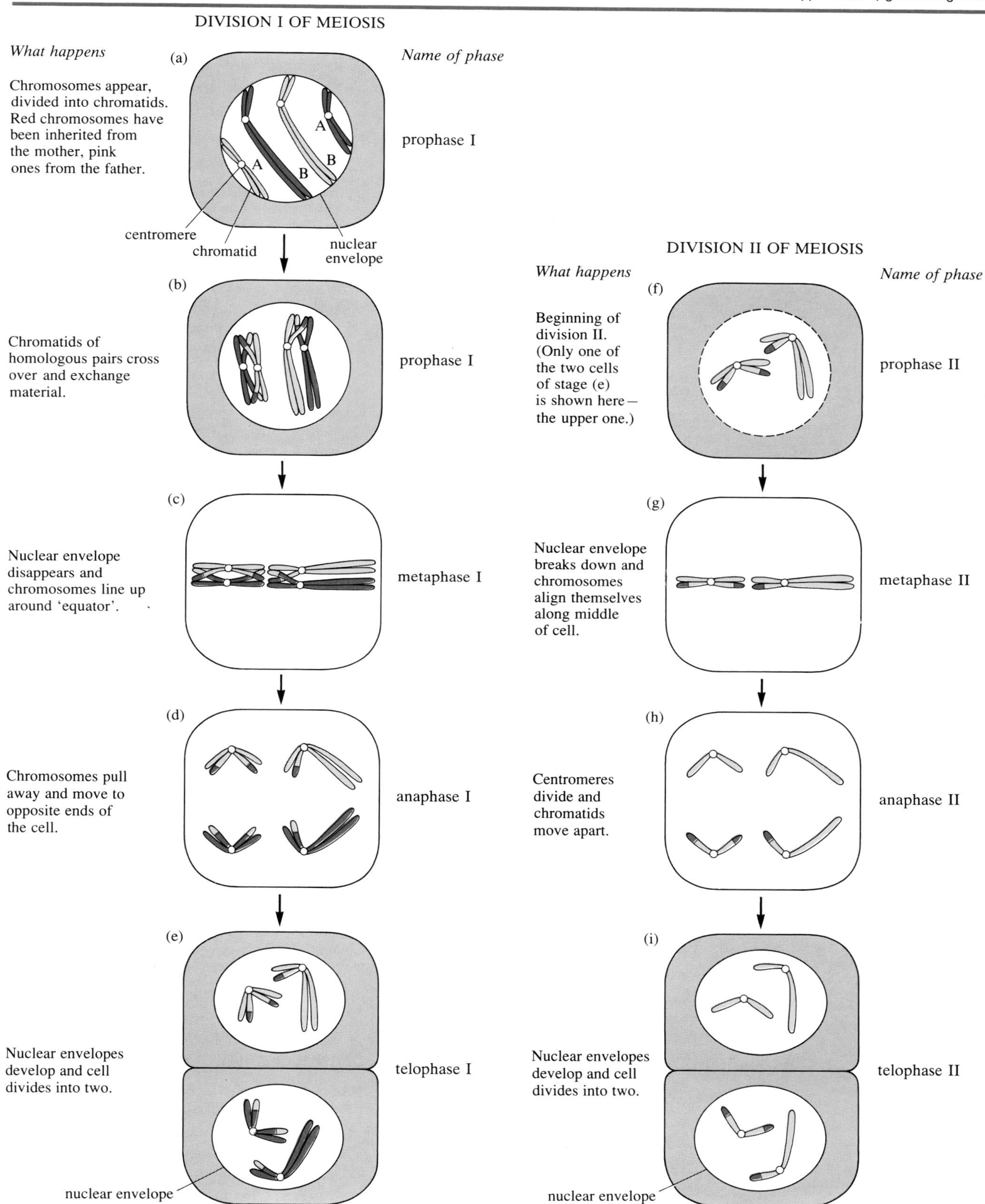

Figure 9.5 The successive stages of meiosis. The stages have been lined up so that equivalent phases of divisions I and II are side by side. Note how meiosis results in halving of the number of chromosomes and in the partial exchange of genetic material between homologous chromosomes.

comes from the mother, the other member from the father. The two sets come together in the new individual (the zygote), when the germ cells (the gametes) of the parents fuse at fertilization. This means that the germ cells must carry only half the final number of chromosomes (23 in humans), otherwise each successive generation would have twice as many chromosomes as the previous one. To achieve this, germ cells have to undergo a *reduction* division, or meiosis, in which the number of chromosomes is halved (see Figure 9.5). The halving is not done simply numerically; each gamete must carry a *complete* set of the chromosomes, and hence genes.

Just which member of each of the parent's pairs of chromosomes goes into each gamete is completely random. Moreover, there exists the opportunity for exchanging genes between members of a pair during the course of meiosis—via the process called *crossing over* (see Box 2.1 and Figure 9.5). This ensures that no gamete is the same as any other. This is why, although we resemble our parents, and our brothers and sisters, and, indeed, have many of the same alleles as each of them, we are not identical to them (except in the case of monozygous twins, which result from the splitting of a single zygote).

How similar are we to our parents and siblings? Half our genes come from each parent, therefore, for *any one gene,* there is a one in two chance that it will have come from our mother, and a one in two chance that it will have come from our father. Likewise, for any one gene, there is a one in two chance that our sibling inherited the same allele (i.e. the gene from the same parent) as we ourselves did. So, on average, siblings have half their genes in common. First cousins have a quarter of their genes in common—their mothers or fathers who were siblings have half *their* genes in common, and the first cousins have each inherited half of these: half × half = quarter.

These rules of inheritance are based on the results of strictly controlled breeding tests carried out on organisms whose alleles are known. For example, Gregor Mendel, whose experiments first established the rules, worked on characters in the pea. He looked at genes which only had two alleles, one dominant and one recessive, and made pure-breeding, i.e. homozygous, strains of each type (see Figure 9.6). When he crossed the two strains, the offspring (called the **first filial generation**, or $\mathbf{F_1}$) all resembled *phenotypically* the parent carrying the dominant allele, because they each carried one dominant allele, and one recessive allele. But when Mendel crossed the F_1 plants among themselves, the offspring included some with the recessive phenotype. This is because, when the gametes fused, some carrying one recessive allele fused with others also carrying a recessive allele. This made the zygote *homozygous recessive,* so the recessive phenotype was evident. On average, this happened once in every four fusions: the homozygous recessive type has a one in four chance of appearing in this cross. The other three-quarters of the offspring all had the dominant phenotype, but in reality they fell into two classes: those which were homozygous dominant (one in four) and those that were heterozygous (two in four, or one in two). ■

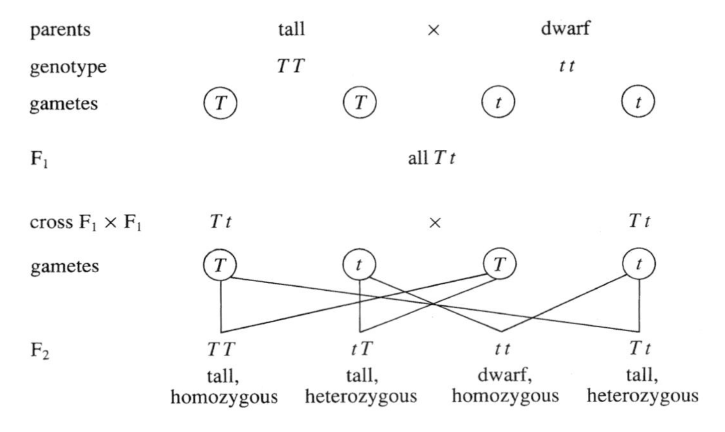

Figure 9.6 A simple genetic cross in pea plants, involving one pair of alleles: *T* (dominant) and *t* (recessive).

More than 3000 diseases are now known to be the result of mutations in single genes. Some of these are extremely rare, but others are relatively common, like cystic fibrosis, which affects one in 2500 newborns (see Section 9.4, later). Some of these diseases are the result of dominant mutations, and are therefore always seen when the abnormal allele is present. Others, however, are the result of a recessive mutation.

▷ What does this imply about the genetic make-up of sufferers of such diseases?

▶ They must carry *two* copies of the mutant allele, i.e. be homozygous recessive (Box 9.1).

It is often the case that affected children are born to parents with normal phenotypes.

▷ What does this imply about the genetic make-up of both parents? (*Note*: sometimes this is the result of a new mutation, as you will see below; for the moment, however, assume that no new mutation has arisen.)

▶ They are *heterozygous* for the mutant allele, i.e. each parent is a carrier of the recessive allele, although it is masked by the normal, dominant allele.

One example of an occasion where gene probes can make a huge difference to patients is in the case of Huntington's disease. This disease is characterized by a degeneration of brain tissue, causing the development of continuous involuntary movements and progressive mental deterioration. Huntington's disease (HD) is due to the presence of an abnormal dominant allele. As we implied above, diagnosis of a disease caused by a dominant allele is relatively straightforward, because the disease phenotype is not masked. However, there is a problem with HD: the symptoms do not usually develop until middle age, after the allele may have been passed on to children. The children are in an unenviable situation: if we assume that just one parent is hetero-zygous for this dominant allele, then each child has a one in two chance of inheriting the allele, and therefore succumbing to the disease. If they themselves have children, these too will be in the same situation. People who do not inherit the allele have no chance of passing it on to their children, and so can live without this worry; those who do inherit it can at least make preparations for infirmity, and will be aware of the risk of passing the allele on to their children. There is now a gene probe which will specifically detect the HD allele. Children whose parents develop HD can now be tested, with luck before they themselves have had time to start a family.

When there is a family history of a genetic disease resulting from the presence of two copies of an abnormal *recessive* allele, it is very important for prospective parents to know whether they are carriers of the allele—unlike the situation in HD, where at least the risk factor will eventually be known, symptoms of these diseases will never develop in carriers (heterozygous). If the prospective parents are not carriers, or if only one of them is, then the risk of having an affected child recedes to the level for the general population.

▷ What does this level depend on?

▶ The mutation rate for that gene in the general population.

▷ If both prospective parents are carriers of a recessive disease-causing allele, what is the chance that their child will be affected?

▶ One in four. This is the frequency with which the two recessive alleles carried by the parents will come together in the zygote (Figure 9.6).

Depending upon many factors, including the particular disease involved and the ethical convictions of the parents, if either or both of them is found to carry a disease-causing allele, the decision may be made not to have children. But, equally, the parents may decide to go ahead. **Prenatal diagnosis** would allow them to know whether their child would suffer the disease, and an early termination of pregnancy, or appropriate preparations for bringing up a sick child, would be options.

▷ Genetic engineering might allow another option. What is it?

▶ One option might be to fertilize eggs *in vitro,* screen the resulting embryos with a gene probe that specifically recognized the mutant, disease-causing allele, and re-implant only those embryos that were carrying the normal allele.

This kind of embryo screening represents the frontiers of what is technically feasible at present. Many other sorts of prenatal diagnosis are already made. Some can be made from the appearance of the foetus in ultrasound scans. Other tests can be conducted on cells taken from the growing foetus—either by *amniocentesis* (removal of some of the amniotic fluid, which bathes the foetus and into which some of its cells are shed); or by *chorionic villus sampling* (removal of cells from the placenta). Both these techniques, particularly the latter, carry a risk of causing an abortion, so unless the disease to be tested for is a serious one, the techniques are not carried out. However, given that foetal cells can be obtained, usually without harm to the foetus, it might be a simple matter to test the foetal DNA for any known genetic diseases. This has the advantage over embryo screening that it involves fewer laboratory manipulations *(in vitro* fertilization techniques, although now quite commonplace, are still technically tricky), but it has the disadvantage that, at the time of diagnosis, the mother will already be pregnant. This raises ethical, and emotional, questions.

These are early days for prenatal diagnosis, and few disease-causing alleles have been identified yet; far less have diagnostic gene probes been produced for them. Those which are currently available are shown in Table 9.2. (This table includes probes for non-inherited conditions too.) Note that these probes are reacting with *mutated* DNA; they are not appropriate for diseases caused by under- or over-expression of a normal gene, such as in the case of certain types of sugar diabetes. Because the probes are for specific sequences, they will pick up not only abnormal alleles (i.e. mutations) which have been passed on from the parents, but also any mutations which have newly occurred. This might be important for diseases (such as cystic fibrosis, see Section 9.4) where a high proportion of cases arise from new mutations in each generation.

Table 9.2 Diseases for which gene probes are available and an indication whether gene therapy is currently a possibility. (Adapted from Anderson, W. F., 1992, *Science*, **256**, pp. 808–813.)

Disease	Gene therapy	Inherited
malignant melanoma	yes	no
paediatric acute myelogenous leukaemia	no	no
neuroblastoma	no	no
chronic myelogenous leukaemia	no	no
adult acute myelogenous leukaemia	no	no
adult acute lymphoblastic leukaemia	no	no
renal cell cancer	no	no
cystic fibrosis	yes	yes
Huntington's disease	no	yes
malaria	no	no
haemophilia (some forms)	yes	yes

Gene probes are actually applied to the fragments of DNA obtained from a restriction enzyme digest of the DNA from the cells of a patient (prospective parent or foetus) (Sections 4.3 and 6.4).

▷ Why is it necessary to cut the DNA before testing it?

▶ Simply because otherwise it would be too long, and the DNA strands would be so tangled that the probe might not be able to gain access to the sequence being sought. This could give a false negative result.

If the sequence recognized by the probe is present, then the probe will bind to it. If the probe is readily identifiable (for example because it has previously been radioactively labelled), then it can be detected among the restriction fragments after washing thoroughly to remove unbound probe. This would be clear evidence that the sequence, characteristic of a specific disease, is present in that patient's DNA. The reaction is highly specific, although at present the limits of resolution demand that the test be performed on a relatively large number of cells. This makes gene probes difficult to use for prenatal diagnosis, where, by definition, there are not many cells available to test, but there is no problem about applying them to prospective parents who are willing to donate a few millilitres of blood. With technical innovations, the number of cells required for the test is becoming fewer and fewer, and this could lead to more widespread use of the probes prenatally. The main stumbling block to the use of gene probes is our present level of knowledge about the genes involved in genetic diseases. In many cases, the only evidence for a genetic involvement in a disease is its familial incidence. For some such diseases, the prospect of identifying and locating the culprit alleles is very distant, and the acquisition of enough base sequence information to make a gene probe is even further away. Unfortunately for the sufferers, the incidence of many genetic diseases is so low, that the commercial incentive for researching them is virtually non-existent, particularly with current worldwide economic conditions. However, you will see later (Section 9.4) how gene probes *have* been made and used successfully in the case of cystic fibrosis.

Detection of cancers

You can see from Table 9.2 that most of the diagnostic gene probes listed are for cancers of one sort or another. Cancers are characterized by an uncontrolled proliferation of particular cells, and there is generally no difficulty in obtaining enough cells to test with a gene probe. In the leukaemias, for example, there is an excess of leukocytes, or white blood cells—'leukaemia' means white blood. This is clearly useful for diagnosing a disease which is already in progress. But might gene probes be able to detect cancerous cells *before* the life-threatening proliferation and spread begins?

Although this is by no means always the case, there are some cancers which are associated with particular DNA sequences. The classic example of this is Burkitt's lymphoma, a type of cancer which is common in Africa and the Far East. Burkitt's lymphoma is associated with the presence in cells of a virus, called the Epstein–Barr virus. Thus, in principle, a gene probe for sequences in the Epstein–Barr virus might be used to screen the population at risk *before* the lymphoma developed, and appropriate therapy could be administered to those people who carried the virus, and would therefore probably succumb to the disease. There is now a vaccine produced by genetic engineering (see Section 9.3.1) which can protect against the Epstein–Barr virus. If this vaccine becomes widely available, then screening should not be necessary.

There are now several human cancers which are known to be associated with viral DNA base sequences. In time, probes may become available to test for these.

9.2.3 The Human Genome Project

Genetic manipulations and techniques for determining the base sequence of DNA have now become so commonplace that it has recently been proposed to find out the sequence of the entire human genome—a matter of around 2.9×10^9 base pairs. This is an enormous undertaking, and is estimated to take 15 years and cost $3 billion to complete. The project represents a truly global effort, with laboratories in Europe, the USA and Japan all investigating different chromosomes. It seems obvious that the more that is known about genetic organization in normal individuals, the better will be understood the basis of many genetic diseases. In a way, the **Human Genome Project (HUGO)** is a logical extension of gene probe research: it represents the *ultimate* library for gene sequences. When the work is completed, it will be possible to know the sequence, and hence perhaps the function, of every single human gene. This will allow us to understand not only how DNA base sequence relates to protein structure and hence to function, but also perhaps how the integrated genome works to produce an individual. We should be able to understand why only some alleles are permitted, and others are lethal (see Box 9.1). Philosophically, it could be argued that HUGO represents the ultimate human quest. Intellectually, it seems inevitable that given the technology to carry out the work, it will eventually be done.

However, already two problems have raised their heads. One relates to patenting. Although it has been established that living organisms, and even DNA sequences, are patentable (see Section 8.7), at what point should patentability stop? One of the laboratories involved in the Human Genome Project has been determining the base sequence of short fragments of human DNA, using an automatic sequencer, at the rate of 168 fragments a day. The laboratory has applied for patents on these sequences, even though nothing at all is known about their function! (You will learn more about this in Activity 9.1.) This has horrified the scientific community, which, in a leading journal, has criticized the application as a 'land grab, a pre-emptive strike that would promote a worldwide stampede to garner patents on essentially meaningless pieces of DNA' (*Science*, 1992, **225**). This problem shifted the goalposts for the Human Genome Project, and instead of the spirit of international cooperation essential for the completion of the project, the mood looked as if it might become one of secrecy and conflict. There remains the question, first, of whether human DNA base sequences *should* be patented at all. If it is decided that they should, then who should be the patent holder: should it be the financial backer of the sequencing project, the scientists who carried out the sequencing work, humankind as a whole, or the individual(s) whose DNA is sequenced? No doubt you have your own opinion on this.

Activity 9.1 *You should spend up to 30 minutes on this activity.*

Read Extract 9.1. Summarize the main points of the patenting story in about 100 words. Then summarize the different opinions held by the major protagonists—the NIH and the biotechnology industry—using not more than about 100 words for each.

Note: in the extract, the term 'fragments' (of genes) is used to mean restriction enzyme fragments, and so not necessarily parts of *genes*.

The other problem raised by HUGO relates to employment and insurance. As we have said, the more genetic knowledge that emerges, the more becomes known about hereditary diseases—not just what a person *is* suffering from, but also what they are *likely* to suffer from in years to come. (Of course, it is by no means *certain* that people will develop a particular disease; the presence of a particular disease-associated gene represents only a finite probability that the disease will arise at some time, and

Extract 9.1 From *The Economist*, 3 October 1992.

Intellectual property
Obviously not

In applying for patents on over 2 000 genes and fragments of genes, America's National Institutes of Health (NIH) has kicked up quite a fuss. Academic scientists began to worry that concerns over intellectual property would dam the free flow of information. Industry, too, became fearful and confused. Now the patent office has rejected the NIH's claims on a number of grounds, but that does not mean the fuss is over. Far from it.

Early on, DNA research involved small teams of scientists devoted to finding and studying specific genes and the proteins they described, and the patent system could keep up. Now the pace has quickened, and some 16 000 biotechnology patent applications are languishing in a queue. And the genome programme, with its implicit suggestion that describing genes could be an end in itself, and its new technologies for speeding up genetic research, could produce a tail-back at the patent office that overwhelms today's backlog. It is now quite easy to amass piles of genetic data; trawl nets are available where before there was only hook and line.

It was by trawling through existing libraries of DNA that Craig Venter, then of the NIH, now of the Institute for Genomic Research in Gaithersburg, Maryland, produced descriptions of over 2 000 genes and fragments of genes used in the brain. Nobody knows what the genes these fragments come from do, but at least some of them could be the blueprints for proteins that might usefully be made into drugs. As such, they could improve public health and provide companies with profitable products—both things the NIH is meant to do.

So when Dr Venter came up with his first batch of genes and fragments last year and wanted to publish them, the NIH insisted on applying for patents at the same time. Patents have, over the past decade, become increasingly popular in America as tools for getting public-sector innovations into private-sector hands.

In justifying the decision, Bernadine Healy, the agency's director, points to the risks of doing nothing. If the NIH published the details of its fragments without applying for patents, a company which took some of the research further might find itself unable to protect the fruits of this labour. With the original NIH work in the public domain, further studies on any of the genes involved would be 'obvious'—and patents are not given for obvious things. With patents, the NIH could use licensing agreements to encourage new developments. Without them, the whole field might prove unprotectable.

The biotechnology industry is in two minds about such help. Embodying its split personality at last week's Senate hearings on the matter was David Beier of Genentech. He was representing both the Association of Biotechnology Companies (stated position: 'ABC supports the government in its decision to file the Venter patent applications') and the Industrial Biotechnology Association (stated position: 'NIH should not seek patents on either partial genes or gene sequences whose biological function is unknown.'). Some biotechnologists, Dr Beier said, worry about the costs and capriciousness licensing might add to the already tortuous process of product development. Others worry that, if the NIH does not patent the discoveries, its employees will—as the law allows them to.

Now this confusion has been further complicated by the patent office's preliminary findings. Dr Healy discussed their gist at last week's hearing. The patent office found that the gene fragments were not novel, because they were taken from existing DNA libraries; they were not useful, because although the fragments allow you to find complete genes (which the NIH had argued was use enough), the genes' functions are unknown. The one attribute the patent office did find was obviousness. So that was three strikes against the NIH applications.

Dr Healy is particularly worried about obviousness. Apparently the patent office claimed that some genes Dr Venter had described in full were not patentable because descriptions of small portions of them had been published elsewhere. In a sense this justifies the NIH's original position that prior publication of fragments can stand in the way of patents for later full descriptions; it suggests that, as more and more snippets are published, it may become harder and harder to patent anything worthwhile. But that does not mean the NIH's solution would work even if its patents were granted. It is far from clear that a patent on a fragment could be extended to cover a full description of a gene.

The NIH will probably pursue the matter, and it seems certain that other applicants are also looking for protection of similar results more discretely. If patents are awarded, many fear a 'gold-rush' to the patent office which will slow research and development. If the patent office remains unconvinced, then it may become hard to patent any gene at all: the march of technology may yet make everything obvious.

other factors may stimulate or prevent the disease from actually occurring.) The Association of British Insurers predicts that in the not too distant future genetic tests will be available, costing around £5 each, which will indicate the individual's susceptibility to a wide range of conditions. Insurance is based on uncertainties and probabilities. Although actuaries can predict how many people will die young, they have not yet been able to predict the fate, and hence the insurance premiums, of individuals. But all this may change. The problem is likely to apply to employment too—should individuals with a *predisposition* to, for example, lung cancer, be allowed to work in the mining, or fibre, industries? If not, will this lead to a stratification of the working population into those who can do 'dirty' jobs, and those who can't? There are calls for legislation in Britain (such legislation already exists in California) to prevent discrimination on the basis of an individual's genes. But if it does become possible for all an individual's genes to be scrutinized, is it likely that insurance companies would forego the privilege? And do we *want* to know what we are likely to die of?

Activity 9.2 *You should spend up to 15 minutes on this activity.*

John X, aged 35, is a factory worker. Suppose genetic screening tests revealed that he is sensitive to machine oil, which can cause an allergic reaction and, sometimes, cancer. Should this information be made available to his employers? Summarize, in not more than 220 words, the arguments for and against doing this.

9.3 Therapeutics

While diagnosis is one side of the medical coin, prevention and cure are the other. The therapeutics sector is enormous: worldwide drug sales were $170 billion in 1990. Drugs are expensive and time-consuming to produce (Section 9.1); it is estimated that to develop a new drug costs in the region of $200 million, so, as you have seen, drugs must be priced high to offset these costs. Two drugs produced by rDNA technology reached the top 50 sellers in 1990: Humulin, Eli Lilly's human insulin produced in genetically engineered bacteria for the treatment of diabetes, and Epogen, Amgen's trade name for erythropoietin, a substance that promotes red blood cell production and is used to combat anaemia associated with kidney failure. It is reckoned that the specificity of biological products and the low cost of production once the system has been worked out will ensure that genetically engineered therapeutics play an increasing part in the pharmaceutical companies' portfolios. There are three major areas where genetic engineering is likely to have a large impact: vaccine production, new or improved therapeutic drugs, and a completely new option, gene therapy. We examine these below.

9.3.1 Vaccines

Vaccines are preventive measures against disease rather than cures, but they are included in this section because they are products which are *administered internally* to patients, and which are therefore subject to the same regulatory controls as drugs.

The concept of vaccination is not new. In 1796 Jenner examined the phenomenon whereby milkmaids who had suffered from the relatively minor cowpox appeared not to get smallpox, a scourge of the time. He made a crude preparation of pus from cowpox sores, which he rubbed into scratches on the skin of volunteers. The volunteers were subsequently also resistant to smallpox. The virus which gives rise to cowpox is called *Vaccinia*. The technique of **vaccination** involves exposing the body (and thus the immune system, see Section 9.2.1) to a 'mild form' of a pathogenic microbe, in the knowledge that the body will then be able to combat a subsequent attack by a more virulent form of the pathogen, or a close relative of it. In this country, all children are offered vaccination against common diseases, and, as you will know if you have visited certain foreign countries, many other diseases can also be avoided by vaccination. Vaccination protects not only against sickness and death, but also against disease transmission, and smallpox, target of the first vaccine, has now been eradicated. However, globally, the picture is less rosy. Millions of people, mainly children, die each year from diseases which could be prevented by appropriate vaccination. For example, measles, which we consider a fairly minor childhood illness, and against which, in any case, our children can be vaccinated, affects 67 million people, and causes two million deaths *each year*. Wider use of the measles vaccine, and of existing cheap vaccines against polio, tetanus, whooping cough and tuberculosis could save three million child deaths a year. As yet there are no vaccines against the major tropical diseases malaria, schistosomiasis, or Chagas' disease, which together account for around two million deaths per year. (The deaths are the tip of the iceberg: malaria affects more than 100 million people per year; Chagas' disease, 16–18 million; and schistosomiasis, a staggering 200 million.) It is estimated that for a vaccine to be affordable by the developing world, it has to be priced at about $1 a dose. Given the high cost of drug development, and the long development time (as much as ten years, as we mentioned above), many drug companies are reluctant to get involved in this area, although the World Health Organization and other bodies fund a considerable amount of vaccine research.

Nowadays, vaccines are no longer derived from the pus of infected individuals, but the principle remains the same. The 'mild form' of the pathogen is generally either killed organisms, or ones that have been treated to make them harmless—so-called *attenuated* strains. More recently, a third type of vaccine has appeared: *subunit vaccines* (see below). Bacteria and viruses are coated with an arrangement of proteins, which protect them and allow them to attach to the cells which they are infecting. These *coat proteins* are obviously foreign to the healthy human body, and, as you have seen, foreign proteins act as antigens, which can be acted on by the immune system. In fact, the coat proteins are the only part of the disease-causing organism that the immune system 'sees'—once the organism is inside the cell it is protected from antibody attack. This is why parasitic diseases, such as malaria, in which the pathogen lives almost entirely within the host's cells, are so hard to prevent. A good approach to vaccination, then, is to expose the patient *just* to the coat proteins of the pathogen; after all, this is all that will be seen on subsequent exposure. Such vaccines are called **subunit vaccines.** Their use removes the slight risk, inherent in all conventional vaccines, that the patient will actually suffer a dose of the disease itself due to the presence in the vaccine of occasional 'full-strength' pathogens.

▷ Can you see how genetic engineering can help to make subunit vaccines?

▶ By isolating the gene for the pathogen's coat protein, inserting it into a laboratory organism, and getting it synthesized on a large scale.

Besides being an easier way to produce vaccines, this approach also means that the process of vaccine production is much safer, as there is no risk whatsoever of the pathogen escaping. One vaccine recently made in this way is that against hepatitis B, whose coat proteins were synthesized in bulk by genetically engineered yeast cells. Currently the price of this vaccine is about $15 a dose, so costs will still need to come down before it can be made available outside the developed countries.

One big advantage of recombinant DNA approaches to vaccine production is that the coat proteins themselves can be intensively studied to discover which parts are the most likely to elicit an immune response. Many of the most antigenic proteins have carbohydrates attached to them, so it may be possible to add suitable carbohydrate residues to the coat proteins to stimulate the immune system still more. Furthermore, for diseases involving many different antigens (such as parasitic diseases where different stages in the parasite's life cycle present different antigens), various antigenic subunits could be tacked together to make a *multiple vaccine*. The whole system is so flexible, that many possibilities are available to vaccine makers at little more than the original cost.

One drawback to using subunit vaccines is that sometimes the isolated coat proteins are not as good at stimulating the immune system as are intact pathogens, possibly because they need to be held in a particular configuration by the rest of the microbe. One way around this is to put the pathogen's coat protein genes into the DNA of a *harmless* microbe, and then use the whole engineered microbe as a vaccine. The proteins will be presented in the correct configuration on the microbe's surface, and the vaccine will therefore be a good challenge to the immune system. But because of the absence of disease-causing genes it will be incapable of causing any disease. *Vaccinia* virus (the cowpox virus) has been used in this way to generate immunity to rabies. Coat protein genes were isolated from the rabies virus, and inserted into *Vaccinia* virus, which expressed the genes so that it carried rabies coat proteins as well as *Vaccinia* ones. Other viruses whose coat proteins have been inserted into *Vaccinia* include herpes simplex and influenza. However, there is a reluctance to use *Vaccinia* in humans, just in case it should mutate to a smallpox-like virus. (Remember, the two are very similar — this was the basis of Jenner's experiments. Clearly, our expectations of vaccine safety have increased with time!) If these experiments could be repeated using a different 'host' virus instead of *Vaccinia*, this approach might prove fruitful.

An alternative approach to producing vaccines by genetic engineering is that the pathogen itself could be engineered to remove the disease-causing genes, while leaving the rest intact. This could then be used directly as a vaccine, because the coat proteins would have the correct configuration, but the absence of disease-causing genes would remove any risk of the disease itself. One vaccine of this type now on the market is against whooping cough. The pathogen involved, a bacterium, *Bordetella pertussis*, has had just two codons modified in the gene for pertussis toxin, which causes the disease, to produce a non-toxic form of the toxin. Many vaccines made by this kind of rDNA technology are currently under development, including ones for herpes simplex, malaria and AIDS. Vaccines against some animal diseases are also under development; viral diseases such as foot and mouth disease cause enormous losses to world agriculture.

9.3.2 Drugs from bugs

When prevention of a disease is impossible, then a cure must be sought. Over the last 50 years, the major advance in this area has been the introduction of antibiotics, which have proved to be remarkably effective against bacterial infections. Antibiotics are produced, in microbes, by very complex metabolic routes, and are therefore not necessarily readily amenable to genetic engineering.

▷ Why is this?

▶ Each step in a metabolic pathway is catalysed by one enzyme, which is encoded by one gene. So a complete metabolic pathway is the product of *many* genes, all or most of which might need to be transferred by genetic engineering. This might be very difficult, given current technology (Section 7.4).

But many diseases are not the result of bacterial infections, and so require other means of treatment. It is in these areas that genetic engineering can make, and indeed has already made, a spectacular impact. Because of our increasing understanding of how the body works, we are frequently able to identify the system, and sometimes even individual substances, which are at fault in a variety of medical conditions. Therapy can therefore be targeted with great specificity, with the design of 'rational' drugs. In many cases these are drugs aimed at helping the body's own defence mechanisms to overcome the problem. Examples of this are interferon and tumour necrosis factor (see below). In all cases where genetic engineering has made a contribution to drug production, the drug is the product of only *one* or, at most, *two* genes: unlike antibiotics, drugs such as these are relatively easily made by rDNA technology.

Human insulin

One of the first such drugs to be produced by genetic engineering was **human insulin**. In the past, insulin for the treatment of sugar diabetes was extracted mainly from pig pancreases and purified. Over a lifetime's exposure, many diabetics developed antibodies to the pig 'labels' on the insulin, that bound to the insulin and made it ineffective. There was therefore a *prima facie* case for using human insulin, which would not be recognized as foreign. However, a sufficiently large supply of human pancreases was not available. So when the technology became available, the human insulin gene was isolated and inserted into bacteria, where human insulin was successfully produced. Insulin consists of two polypeptide chains, held together by disulphide bridges. But the two insulin chains actually derive from one original polypeptide chain, from which a section is cut out at a late stage (see Figure 9.7). Since bacteria cannot carry out this post-translational modification (see Section 7.2.1), the researchers side-stepped the problem. They began by making two synthetic 'genes' each coding for just one of the chains. These two genes were cloned separately in bacteria. The polypeptide products were isolated and then combined chemically, outside the bacteria. The result was synthetic human insulin, which was given the name 'Humulin' (by Eli Lilly). Humulin is now widely available and was number 39 in the list of top selling drugs in 1991.

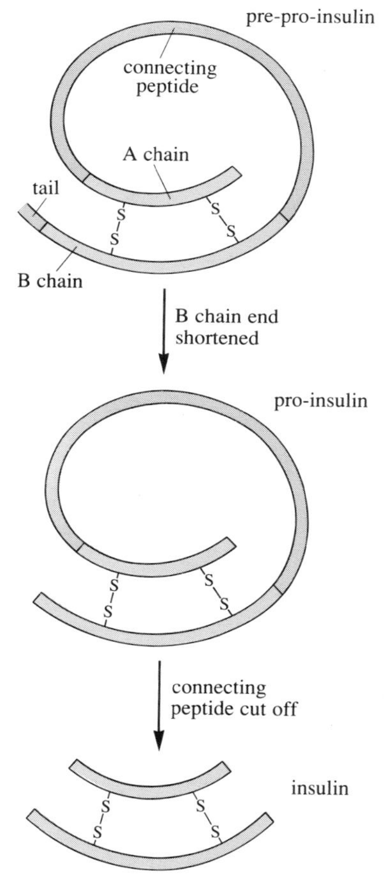

Figure 9.7 The post-translational modification involved in the biosynthesis of insulin. The gene product is the insulin precursor, pre-pro-insulin. Removal of a short 'tail' section produces pro-insulin. Excision of the 'connecting peptide' leaves the mature insulin molecule — A chain and B chain joined via two disulphide bridges.

Many of the rational drugs now at various stages of production are, like insulin, natural products which boost the body's own responses. Some, for example, boost production of blood cells. This can be very important in several cases. Epogen (erythropoietin), mentioned above, promotes red blood cell formation and so is useful for the treatment of anaemia (lack of red blood cells). Other products, such as various growth factors collectively called colony stimulating factors (CSFs), can boost other body functions. Some promote the formation of *white* blood cells, part of the immune system, and are therefore widely used with cancer chemotherapy, which tends to kill dividing cells, including the white blood cell progenitors, leaving the patient open to infection. Other CSFs are used to promote bone marrow growth after transplantation, and as a general pick-me-up in the treatment of AIDS, which is characterized by a loss of immune functions.

Another drug produced by rDNA technology that has had a considerable life-saving effect is tissue plasminogen activator (t-PA), which acts in the body to dissolve blood clots (Section 7.2.2). t-PA, if given to heart attack victims shortly after their attack, has been shown to improve survival rates dramatically. It is now administered routinely to such patients, and also to those suffering strokes, deep vein thrombosis, and unstable angina. In 1990 t-PA had sales of more than $200 million, and made its parent company, Genentech, the world's number two company making drugs by genetic engineering (number one was Amgen, the company making Epogen). (*Note*: strictly speaking, this is not a 'drug from a bug', as mammalian tissue culture cells, suitably engineered, are used.)

Table 9.3 Drugs made by genetic engineering. Note that some of these products are still undergoing trials and are not yet commercially available.

Name	Condition treated
Humulin	insulin-dependent diabetes
Ceredase	Gaucher's disease
deoxyribonuclease	cystic fibrosis
t-PA	blood clots
Factor VIII	haemophilia
CSFs	blood cell deficiency
human growth hormone	dwarfism
α-interferon	Kaposi's sarcoma
superoxide dismutase	oxygen toxicity in premature babies
interleukin	some cancers
tumour necrosis factor	cancer
monophosphoryl lipid A	septic shock

Interferon

There are now a number of drugs which have been made widely available through genetic engineering. A few are shown in Table 9.3. But perhaps one of the most tantalizing is **interferon**. Interferon was discovered in 1957 by virologists Isaacs and Lindenmann, working in London. Interferon was so called because it was the agent that appeared to cause a phenomenon known as viral interference, in which people suffering from one kind of viral infection never suffered from another simultaneously. It seems that cells under attack by a virus produce a substance which attacks *all* viruses. However, interferon is species-specific to the infected organism: only interferon produced by human cells is active in human bodies. The potential for a general anti-viral agent is clearly enormous: even today there are very few drugs that are effective against viral infections. Interferon is produced only by cells which have been exposed to attack by a virus (i.e. challenged). The first 'use' of interferon came soon after its discovery, when it was spotted in 1960 by *Flash Gordon* (see Figure 9.8), but trials in the real world were hampered by low availability: challenged cells produce only infinitesimal amounts—interferon is so biologically active that not much is needed. Isaacs and Lindenmann were never able to purify enough interferon to determine its chemical nature; it is now known to be a protein.

Once interferon had been discovered, and its anti-viral effects had started to be appreciated, other researchers joined the fray and tried to devise ways of obtaining larger amounts of the elusive substance. The main producer of interferon, a Finnish researcher called Kari Cantell, used white blood cells derived from the 800 pints of

Figure 9.8 The first clinical use of interferon is shown in this 1960 *Flash Gordon* comic strip as space medics inject it to save the victim of an extraterrestrial virus.

blood donated daily to the Finnish Red Cross. Cantell's method was to isolate the white blood cells and culture them in the laboratory. The cells were challenged by a virus, and cultured for a further 24 hours, after which the interferon was purified. The purification was a frustrating process, as 99% of the interferon itself was lost during the process, and what was left was very impure—not really of a quality suitable for injection into patients. By 1979, Cantell had improved the process to the stage where he produced 400 billion units of interferon per year. (A unit of interferon is defined in terms of its activity in tissue culture; the daily dose for a patient was around 15 million units.) This was just about enough for the action of interferon to be studied more rigorously, although it was very expensive: a daily dose cost several hundred pounds. However, because of the difficulties associated with working on interferon, research interest had waned, and further progress ceased. But interferon regained its popularity when it was found that, in addition to its anti-viral effect, it might also be active against certain kinds of cancer. Research took off again, funded by money for cancer research.

Quite a lot is now known about the nature and action of interferon. Much of this information has been gleaned as the result of harnessing the techniques of genetic engineering to the problems of availability and purity.

▷ How could genetic engineering help to solve these problems?

▶ If the interferon gene could be cloned in, say, bacteria, then interferon could be produced in large quantities. There would be enough to isolate it easily from the bacterial culture medium, and to purify it cleanly.

Sure enough, the main step forward in interferon research came when interferon genes were cloned in *E. coli*, allowing the easy production of milligram quantities. In fact, the cloning work helped to establish that there are actually *three* main groups of interferons (see Table 9.4); some subtypes also exist. Different types appeared to be active against different conditions. More than a dozen interferon genes have now been cloned.

Table 9.4 Types of interferon and sites of production.

Name	Major producing cell type
α	leukocytes, a type of white blood cell
β	fibroblasts, a type of cell involved in producing connective tissue
γ	macrophages, another type of white blood cell

The high yield of interferon produced by genetic engineering (typically 10^9 units per litre of culture medium) meant that the door was opened for properly designed clinical trials of the substance, to test both anti-viral and anti-cancer properties. To date, the results have been largely ambiguous, with some known viral diseases, such as the common cold, apparently unaffected by interferon. Furthermore, the side-effects, which include confusion, aching joints, anorexia, seizures and liver toxicity, are bad enough to preclude the use of interferon in all but the most serious cases. This is puzzling, since interferon is a natural product which appears to be well tolerated when produced by one's own cells. One suggestion has been that the precise cocktail of interferon types is important both for efficacy and for tolerance. The observed toxicity in clinical trials remains to be explained. Nevertheless, in 1991, γ-interferon was licensed for use against chronic granulomatous disease (a serious defect in white blood cells, resulting in failure to kill microbial pathogens), and various α-interferons for use against genital warts, hairy cell leukaemia, a type of hepatitis, and the AIDS-related cancer, Kaposi's sarcoma. There are now about a dozen brands of interferon on the market. None of this would have been possible without the coming together of humanitarian and commercial interests and new technology. The interferon story proved that genetic engineering technology *could* work, and *could* in principle make products to revolutionize therapy. It is ironic that, having gone from a situation of having too little interferon to test on patients, we may now be faced with a situation of having too few patients upon whom to test the various interferons!

Mini-antibodies for therapy

We outlined in Section 9.2.1 how antibodies, the complex products of many, rearranged, genes, could, against all the odds, be produced by genetic engineering.

▷ Can you remember how this was achieved?

▶ By isolating and cloning the (rearranged) gene for *one* part of the antibody, to produce a single-domain antibody, or mini-antibody.

Conventional antibodies are already widely used in diagnosis and therapy, but the use of mini-antibodies for therapy is still in its infancy. However, one very promising line of research, pioneered by the Scottish company Scotgen, is the production of a humanized antibody (see Section 9.2.1) against respiratory syncytial virus. This is the major cause of acute respiratory illness in young children and can cause permanent lung damage. The way in which this humanized mini-antibody was produced is complex, and we do not need to go into the precise techniques here. The procedure involves the basic trick of making a cDNA copy of mature mRNA extracted from antibody-producing cells (Figure 9.2).

▷ Why was mature mRNA the starting point?

▶ mRNA is the final transcript, that has already undergone post-transcriptional modification to remove introns. Thus, it represents the coding parts of the gene (Box 5.2).

There is no conventional antibody against respiratory syncytial virus, for use either as a vaccine or even as therapy once the disease has become established, so this mini-antibody holds out hope of being a significant therapy for a serious childhood illness. It has been shown to bind to viruses in samples of phlegm taken from patients, and has both prevented and cured the disease in laboratory animals. Clinical trials will show whether it works this well in humans too.

9.3.3 Gene therapy

One of the most exciting aspects of genetic engineering, and certainly the one which has captured the public imagination, is the prospect of being able to alter an individual's genes. Leaving aside *eugenic* engineering aimed for instance at producing blue-eyed, blond-haired individuals, we will concentrate on the therapeutic use of the technology.

As you saw above, over 3000 known genetic diseases are caused by defects in a *single* gene. There are two approaches to treating this kind of disease. On the one hand it should, in theory at least, be possible to provide some kind of **replacement therapy** by giving the patient quantities of the *gene product* that is lacking. The products could be made (in, say, bacteria) from an intact, non-mutant gene by the sort of rDNA techniques that are now familiar to you.

▷ What kind of molecule would be administered?

▶ A protein. (However, if the defective DNA sequence was a *regulatory sequence with no protein product*, this kind of replacement therapy could not be done.)

Indeed, replacement therapy is already widely used, for example the administration of human growth hormone, produced by rDNA technology, to children who are deficient in that gene, and who would otherwise be of very small stature. This is very effective, and, as it is a natural human product, the side-effects appear to be non-existent (the exception to this rule is interferon, as you saw above). However, one reason why externally administered growth hormone can combat dwarfism is that the hormone has its effects at many sites throughout the body. Administration by almost any route will still allow the hormone to reach all of its targets. But this is certainly not the case for many other genetic diseases. Gaucher's disease, for example, is caused by the lack of an enzyme needed to break down a lipid in body fat cells. The deficient enzyme, alglucerase, (sold as Ceredase; see Table 9.3) can be specifically targeted to the particular cells where it would normally be produced, but without this specific targeting, administration of the enzyme will not cure the disease, because it cannot reach the cells in which its substrate is to be found. Thus it is vital not just to get the replacement protein into the body, but also to get it to specific cell types. Specific targeting is difficult to achieve in most cases, as not enough is understood about the body's 'address labels'. So replacement therapy is of limited use for genetic diseases.

An alternative approach is to use genetic engineering in a rather more fundamental way—to use **gene therapy**. Instead of replacing the protein product of a defective gene, it might be possible to replace the mutant allele itself with the normal, functioning allele. It is important here to distinguish between gene therapy of *somatic* (body) cells, and therapy of *germ line* (reproductive) cells. In the former case, the situation is little different in principle from organ transplantation: only the patient is affected, and any side-effects of the treatment will devolve on the patient alone. But in germ-line therapy, it is a question of manipulating the genes not just of the patient, but also of any future children that the patient might have, and this raises ethical and moral problems.

▷ Which type of therapy, somatic or germ-line, do you think is more open to abuse?

▶ Germ-line therapy. This would allow **eugenics** (the deliberate breeding of a desired type) to be carried out.

Although there are genuine humanitarian reasons for carrying out germ-line therapy—such as to eradicate a debilitating genetic disease from individuals *and* their offspring—the technique could be abused. Perhaps more importantly, it is *perceived* to be open to abuse. So it is not likely to be a widely available therapy, even when it becomes technically feasible—which at present it is not. However, **somatic-cell gene therapy** is an altogether different case. The 'worst-case' scenario for this type of gene therapy is that it might not work and that the intervention itself might harm the patient still further, such as by accidentally activating cancer-causing genes. However, no therapy is risk-free (Section 9.1), but if the potential benefits outweigh the possible risks, then most people would agree that the therapy should proceed. For the foreseeable future it is likely that somatic-cell gene therapy will only be used in very severely affected patients.

However, it is no easy matter to achieve successful somatic-cell gene therapy. In the normal individual, different sets of genes are switched on in different cell types and at different times (Box 5.2 and Chapter 8). Ideally, then, we would like to get an intact copy of a defective gene into the appropriate cell type *and under the correct regulation*. In this way it will substitute perfectly for the resident, defective gene, and will be maintained in that cell and all its descendants.

Somatic-cell gene therapy has recently (1990) been carried out, apparently successfully, in the USA at the National Institutes of Health. The first patient was a 4-year-old girl who was suffering from severe combined immune deficiency disease (SCID), caused by a deficiency of the enzyme adenosine deaminase (ADA). This deficiency effectively wipes out the immune system, and sufferers must live in a plastic bubble with a purified air supply, and must have no contact with other people because they are so susceptible to infections. The therapy was carried out in the following way: a sample of the girl's white blood cells was removed, and copies of a normal ADA gene inserted into them via a vector. The cells were cultured briefly to check that the phenotype was normal, suggesting that the replacement genes were stably expressed, and had not become inserted into a chromosome in such a way as to damage any other genes. Then the cells were transfused back into the patient. Because mature white blood cells do not divide, and live for only a short time, the treatment must be repeated at intervals of a few months. So far the treatment has been supplemented by conventional treatment, a replacement therapy which involves weekly injections of the gene product, ADA itself. Since the treatment, the girl has been in good health and attends school normally. No side-effects of the treatment have been detected. Although repeated blood transfusions sound unpleasant, it is likely that they would be perceived preferable to the alternative of life in a plastic bubble.

However, if the ADA genes could be inserted into the white blood cell progenitors (in the bone marrow), then the repeated injections and transfusions might be avoided, because the patient's bone marrow would be able to make 'normal' white cells permanently. The ADA work has shown that it seems possible to transfer a gene safely and effectively into a human patient. There is now evidence from more than 100 monkey-years and 20 patient-years in individuals who have undergone this kind of gene therapy for various diseases. No side-effects have been observed, and no malignancy or other ill-effects have been found. Further work is now progressing at a reasonable rate in the USA and Europe—in fact, the first case of gene therapy in the UK has recently been reported (January 1993)—but application of somatic-cell gene therapy in general has been held up by regulatory hurdles. Although the ethics of the technique are essentially the same as for organ transplantation (see above), opponents claim that this is the thin end of the wedge, and that it is only a matter of time before germ-line gene therapy takes place. Indeed, it has been argued that it is wasteful to keep administering gene therapy to somatic cells in each successive generation when by carrying out germ-line gene therapy the problem could be dealt with once and for all. Legislation to prevent this, or to regulate it if it is to be allowed, is called for.

Since the ADA landmark, somatic-cell gene therapy has been carried out on several other patients, notably some with advanced malignant melanoma, a fatal form of skin cancer. Here the effect is not so much gene therapy as a radical method of drug delivery. The idea is to insert the gene coding for tumour necrosis factor (a kind of natural anti-growth factor that causes cancer cell death; Table 9.3) into normal white blood cells which infiltrate tumours. This should result in the tumour necrosis factor being released where it will do the most good, and, hopefully, destroy the tumour. As was the case with ADA (above), white blood cells are removed from the patient, copies of the tumour necrosis factor gene inserted into them, and the cells replaced into the patient. As always, there are technical problems to be overcome.

▷ Name some of these problems.

▶ To get the gene into the target cells, to get it stably integrated, and to get it expressed in suitable quantities.

It is not yet known how effective in the long term tumour necrosis factor gene therapy will prove to be, nor whether any side-effects will appear.

There are many other proposed targets for somatic-cell gene therapy, including muscular dystrophy, cystic fibrosis (see Section 9.4), familial hypercholesterolaemia (an enzyme deficiency which renders its sufferers likely to die young from heart disease), and, in the long term, genetically-determined neurological diseases. The main difficulty, once the defective allele has been identified and its normal counterpart cloned, is to find the correct chromosomal site for insertion of the gene. Then the correct cells to be treated must be identified. These must be at the site of disease, easily accessible, easy to manipulate *in vitro*, and should ideally contain populations of dividing progenitor cells, so that the technique need be carried out only once. Bone marrow cells satisfy all these criteria, so it is likely that the next successes will involve diseases where treatment of these cells is appropriate.

9.4 Cystic fibrosis

Cystic fibrosis (CF) is the most common genetic disease among Caucasians, affecting 1 in 2500 babies. Over 30000 people have the disease in the USA, as do 25000 in Europe. Victims live for around 30 years, and usually die from respiratory failure brought about by repeated lung infections. There is no cure. A diagnostic marker for the disease is excessive amounts of salt produced in the sweat, with the characteristic lung and pancreatic abnormalities appearing only later; adults are usually infertile. The severity of the disease appears to depend not only on the particular mutation within the gene (several are known), but also on environmental interactions which are not understood at present.

Currently, the management of CF depends largely on trying to prevent the respiratory problems that account for 95% of deaths. To this end chest percussion, encouraging the patient to cough up mucus, is a daily event. Antibiotic treatment of infections does help, as does the administration of drugs to reduce the inflammatory response to infection of the respiratory passages. The inflammation tends to narrow the tubes and make breathing even more difficult for the patient. The inflammatory response also includes the production of copious amounts of mucus, which, sticky enough on its own, is made even more viscous by the presence within it of DNA from dead cells. One approach to treating this has been the administration by aerosol of the enzyme deoxyribonuclease (produced in microbes cheaply and in large quantities through genetic engineering) which will degrade the DNA into small fragments and thereby reduce the viscosity of the mucus. An additional problem is that the mucus tends to be rather dry and difficult to cough up. This is thought to be the result of an abnormality in *sodium* transport which may arise in response to the fundamental problem of abnormal chloride transport (see below). Sodium uptake can be blocked by a drug, amiloride, and patients receiving this do seem to have 'wetter' mucus. The problem of pancreatic insufficiency has been addressed by the proposal to implant 'mini organs' of normal pancreatic cells held in a matrix of inert material such as the textile Gore-Tex. The secretory functions of the cells would not be impeded, but the cells would be afforded some protection from the patient's immune system. At the time of writing (1992), this approach remains a pipe dream.

Although the disease itself has been recognized for many years, it is only recently that information about the fundamental defect has been forthcoming. Family studies suggested that the CF allele is recessive. There are actually several mutations within the gene concerned that can result in the disease. The gene involved was identified after an enormous amount of work involving the analysis of DNA fragments obtained by treating patients' DNA with a battery of restriction enzymes. Eventually, fragments were identified which were present in sufferers of the disease and in carriers, but not generally in other people.

▷ Assuming that the disease had not arisen as the result of a new mutation, what would you predict about the relative amounts of these fragments in patients and in their disease-free parents?

▶ The patients would have twice as much as their parents. This is because they would have two copies of the gene, being homozygous recessive, whereas each parent, a heterozygous carrier of the disease, would have only one gene copy (Box 9.1).

The disease-associated fragments were tracked down to chromosome 7, and the exact position of the gene was finally mapped in 1989. It is a large gene, some 250 000 base pairs long including 26 introns; its mRNA is about 6 500 bases, encoding a mature protein of 1 480 amino acids. Once the gene had been identified, it was relatively easy to shed light on its function. The gene is expressed mainly in cells lining the respiratory tract, pancreas, salivary glands, sweat glands, intestine and reproductive organs. The protein product of the gene, called cystic fibrosis transmembrane conductance regulator (CFTR), is a membrane protein whose normal function appears to be to act as a channel to conduct chloride ions out of cells (see Figure 9.9). In CF this protein no longer functions correctly, and chloride ion transport is disturbed, resulting in the disease characteristics mentioned above.

As you now know, genetic engineering has made significant advances in both the diagnosis and the treatment of CF, quite apart from its contribution in identifying its molecular cause. As far as diagnosis is concerned, there is now a gene probe available for the CF gene which can distinguish between the normal gene and the mutant, disease-causing alleles. This probe enables prospective parents to learn whether or not they are carriers of the disease. It can also be applied *in utero* so that affected children can be detected as early as possible. Because the disease is so common, there are moves to introduce screening of the whole population, to detect carriers. One suggestion has been to screen by means of a mouthwash. The cells obtained in this way could be used for screening with a battery of probes able to detect all known CF mutations. Two carriers of a CF mutation would be expected to produce an affected child with a frequency of 25% (i.e. there is a one in four chance of any of their children having CF). Also, it has been observed that if the parents carry *different* CF mutations, one of which produces a more severe form of the disease than the other, then an affected child inherits the *milder* form of the disease. Such information may be an important factor in a couple's decision whether or not to have children.

Genetic engineering experiments have shown that 70% of cases of CF result from the same mutation, a deletion of three base pairs resulting in the loss of a phenylalanine residue in a region coding for an ATP binding site in the CFTR, of which there are two (see Figure 9.9b). This suggests that the disease results from a defect in the mechanism controlling the opening of the chloride channels. This is supported by physiological experiments which suggest that the mutant protein channel remains closed more of the time than does the wild-type (i.e. normal) channel.

▷ Do you think that replacement therapy, simply injecting CFTR into the bloodstream, would be a suitable treatment for CF?

▶ No. Replacement therapy involves administering the protein product of the normal gene. In the case of CF the protein is a membrane protein, i.e. an integral part of the cells themselves. Unless some mechanism exists for incorporating free proteins into the cell membrane, simply putting CFTR into a patient's bloodstream would not cure the disease.

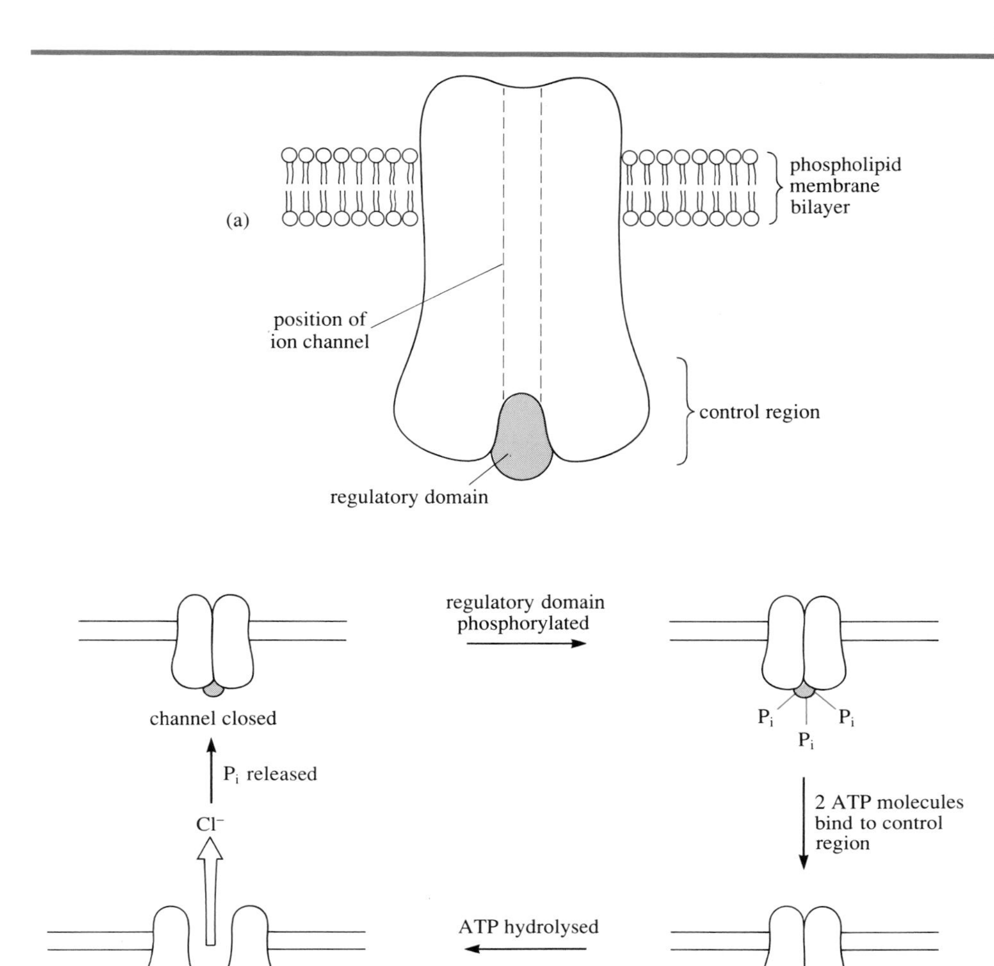

Figure 9.9 (a) Simplified diagram of the structure of CFTR. (b) Proposed mode of action of CFTR in transport of chloride ions out of cells.

On the other hand, the prospect of gene therapy for CF is very attractive. Hopefully, here, the individual with the normal gene now in place would synthesize CFTR in the appropriate cells, *and* insert it into cell membranes. But there are some major hurdles to overcome. It is not really known which cells to treat. Only a small fraction of lung cells appear to produce CFTR in significant quantities, and the chances of treating these cells specifically is remote. It is currently envisaged to administer the normal CF genes (in a suitable vector) to the lungs by aerosol, so it is unlikely that any *specific* targeting could be achieved. However, it turns out that this may not matter. In laboratory experiments it was found that normal CFTR function can be restored when fewer than 10% of the cells have incorporated the normal CF gene. The reason for this is suspected to be due to the way in which the cells communicate with each other. This is by means of *gap junctions,* specialized structures allowing free passage of ions and other materials between cells. Perhaps one normal cell can control the ion flow of many neighbouring cells. Another hurdle relates to the safety aspects. Since the genes are to be applied by an aerosol, it is very likely that escapes into the environment will occur. If copies of CF genes become inserted into cells of individuals other than the patient, is an overproduction of CFTR likely to be dangerous? Early experiments using laboratory animals suggest that it is not. However, much research remains to be done before CF gene therapy becomes the standard treatment for this devastating disease.

Activity 9.3

A hypothetical life-threatening disease called *lethal* is known to have a familial basis. Its pattern of inheritance suggests it is due to a single recessive allele (i.e. only individuals homozygous for the allele exhibit the disease). Symptoms are unpleasant, tend to occur first in early adulthood, and generally prove fatal within 10 years. Following careful studies of DNA from normal and affected individuals, a gene associated with *lethal* is identified. This forms the basis for an *in vitro* diagnostic test using a gene probe for the defective (i.e. *lethal*-causing) allele. Licensing permission is sought to use this test to identify individuals in families with a history of *lethal* who are either heterozygous or homozygous for the *lethal* allele (i.e. carriers of the allele, or prospective sufferers of the disease).

You are a member of a panel set up to consider whether permission to use this test should be granted. What factors should you consider? Give your answer in note form, as a list of numbered points.

Summary of Chapter 9

1 Genetic engineering is likely to make a big impact on the *pharmaceutical industry*.

2 *Drug production* is already heavily laboratory-based, so the adoption of genetic engineering techniques does not represent the major hurdle that it might to other industries.

3 *Diagnostics* can exploit *molecular markers* of disease. Genetic engineering can call upon the *high specificity* of biological recognition to detect such markers.

4 Protein markers can be recognized by *antibodies*. *Monoclonal antibodies*, produced by single clones of spleen cells, are particularly useful for this.

5 Base sequences characteristic of disease can be recognized by *gene probes*. These sequences can be from pathogens, or can be *mutant human genes*.

6 *Prenatal diagnosis* may be possible for some diseases.

7 Cheap, accurate diagnostics raise the possibility of widespread screening, not only for diseases, but also for *susceptibility* to diseases. This may raise *ethical problems*.

8 The *Human Genome Project* is under way to identify the complete base sequence of the human genome. This may increase our understanding of *predisposition to various diseases*.

9 *Vaccination* can prevent individuals from falling prey to specific diseases. Genetic engineering techniques can improve the effectiveness of vaccination by allowing the production of new and better vaccines.

10 Genetic engineering has revolutionized drug production. The first drug to be made by rDNA techniques was *human insulin*.

11 Genetic engineering is particularly suited to obtaining natural substances, normally produced in very small amounts, which can help the body to fight disease using its own, natural, chemicals. The prime example of this is *interferon*.

12 *Mini-antibodies (single-domain antibodies)* can be used for therapy as well as for diagnosis. They can be *humanized* so that they are more effective in the patient's body.

13 Genetic engineering promises completely new approaches to *therapy*, correcting the effects of faulty genes by inserting normal, undamaged genes into the patient's cells.

14 A whole range of new approaches to disease control is being applied to *cystic fibrosis*. This ranges from diagnosis using gene probes to proposed gene therapy using the CFTR gene.

Question 9.1 Antibodies are produced as the result of expressing several different DNA sequences which are rearranged before transcription and translation occur. What does this suggest to you about the feasibility of making antibodies by putting antibody-coding DNA sequences into bacteria?

Question 9.2 A true-breeding white-flowered tobacco plant is cross-fertilized with a true-breeding yellow-flowered one. All the offspring have yellow flowers. Which of the following statements concerning the offspring are true?

(a) The yellow flower characteristic is dominant to the white flower characteristic.

(b) All of the offspring are homozygous dominant.

(c) All of the offspring are homozygous recessive.

(d) If bred with each other, all the offspring of the offspring (i.e. the 'grandchildren' of the original cross—the F_2 generation) would have yellow flowers.

(e) In the F_2 generation, the proportion of yellow-flowered to white-flowered plants would be about 3 : 1.

Question 9.3 Look at Figure 9.4.

(a) Why were there no affected individuals in generation II?

(b) In generation VIII, why were there only four, and not twelve, affected individuals?

Question 9.4 State two advantages and two disadvantages of subunit vaccines.

Question 9.5 What is the most important contribution that genetic engineering has made to interferon research?

10 Agricultural applications of genetic engineering

In this chapter we look at how the techniques of genetic engineering have been — or are planned to be — applied to agriculture. In particular, we examine the application of this technology to whole plants and animals.

10.1 Feeding the world

Agriculture, in the broadest sense, has been responsible for most human nutrition for the last 14 000 years. Ever since we moved from a nomadic, hunter–gatherer way of life to a more settled existence, people have been planting seeds in the knowledge that they would germinate to give plants like those which produced them. Since 12 000 BC there have been many opportunities for refining techniques of conventional plant breeding, and indeed, those techniques have supported a world population which was able to grow to more than two billion by 1940. In the last 50 years agricultural yields have increased 5–10-fold. Around half of this increased yield is due to the emergence of higher-yielding varieties; the other half is the result of more efficient farming practices. Table 10.1 lists the major food crops in the world today.

Table 10.1 The world's major crop plants and numbers of species.

Plant type	Number of species eaten
cereals	8
root crops	3
sugar crops	2
grain legumes	7
oil seeds	7
trees	2
non-leguminous vegetables	15
fruits	15

The 59 species of plant mentioned in Table 10.1 account for most of the world's food production. In 1991 this amounted to more than 7.5 billion tonnes. Most food comes directly from plants; the remainder comes indirectly from plants, via animals. The exact proportions depend on country and culture: in Western Europe and the USA, for example, a lot of meat is consumed. Conversely, in many developing countries meat eating is virtually unknown. This is partly a question of social mores, and partly because it is much more efficient to use land for growing directly consumable crops than for growing animals. A hectare will support five cows but can produce almost 50 tonnes of sorghum per year.

Over the past several years, world food production has been ample for feeding all the world's population. The reasons behind the recent famines are complex, but a major factor is the difficulty of distributing food to remote or war-torn areas. However, as the global population is increasing very rapidly, the problems of feeding people are also set to increase, not least that of producing still more food from a fixed (and actually slightly decreasing — see Section 10.2.1) area of suitable land. In attempts to alleviate at least part of this problem, efforts have sensibly been concentrated on the crops which form the bulk of most people's food intake: the cereals. These are plants such as wheat, rice, maize and sorghum, which are widely grown in different regions of the world. Cereals alone provide 52% of world food in terms of energy.

In the 1960s and 1970s the food production problem was perceived to be so acute that the International Rice Research Institute began a programme of intensive breeding to improve the yields of conventional rice strains and to develop new strains that would be better suited to the inhospitable growing conditions of many of the world's poorest areas. This programme was so successful that the new crop varieties developed have increased world food production by 50 million tonnes per year.

The global population is on target to reach 8–10 billion before the end of this century. This represents an enormous demand for agricultural products. For example, the production of rice, which provides half the total energy intake for two billion people, and 70% of dietary protein in some areas of Asia, would have to increase by 45% this decade. Can present crop plants be grown on a large enough scale to supply adequate nourishment for all these people? And, if not, can conventional breeding techniques produce better varieties? It is generally believed that further progress by conventional plant breeding is likely to be neither rapid nor extensive enough to cope with such demands. Something new and radical is required. Might genetic engineering be able to offer anything here? Clearly, somebody thinks so, as in 1990 nearly $250 million was spent on R & D in plant biotechnology. In this chapter we will look at some of the main targets for genetic engineering in plants, and also in animals. We begin with plants.

10.2 Targets in plant breeding

There are many problems involved in growing plants for food, but they can be assembled into four major groups:

- availability of suitable land
- dependence on fertilizers
- attack by diseases and pests
- competition from weeds.

10.2.1 Land availability

In most populated areas it is generally true to say that most of the land suitable for farming is actually used for this purpose. Marginal land that is not readily available for growing crops, for example because it is too mountainous, can often be used for grazing animals. About 10% of the Earth's land surface is currently used for agricultural purposes, but globally, the land available for farming is diminishing. The recent tendency towards intensive farming, and the clearing of more land by removing trees with extensive root systems which stabilize the soil, coupled with climatic changes, has resulted in a considerable amount of soil erosion and increasing desertification. The interactions are complex, and we cannot go into them here: suffice it to say that more and more land is becoming unsuitable for agriculture, with poor fertility and water-retaining properties. Conventional modern crop plants require very fertile soil (they are *heavy feeders*) and reasonably wet growing conditions, so the areas suitable for growing these are decreasing—not an ideal situation, given the increasing demand for high yields. However, there are plants which are able to thrive in more or less hostile conditions. A well-known example of this is rose-bay willow herb, which is one of the first 'pioneer' plants to colonize inhospitable areas such as slag heaps. Other plants can survive in brackish water which, with a rising sea-level, is encroaching on many coastal farming areas. Yet other plants can survive drought conditions well. Ideally, plant breeders could transfer the 'survival' characteristics from such plants into conventional crops. However, as you know (Chapter 2), conventional breeding cannot generally overcome the species barrier, (although there are exceptions to this rule, such as the wheat–rye hybrid, triticale).

▷ How might genetic engineering help here?

▶ By transferring the genes for, say, drought tolerance into crop plants.

This is fine in theory, but the major stumbling block so far has been to *identify* the specific genes involved. There are two basic problems: first, very little is known at all about the genetics and biochemistry of, say, rose-bay willow herb, and it would be difficult to attempt any kind of gene identification without a considerable amount of background work. All of this would take time (one of the main problems with genetic studies of higher plants is that the generation time is roughly a year). The second problem is that the physiological mechanisms which give rise to drought tolerance or other 'survival' features are often highly complex, involving the products of many different genes. As you have learned (Section 7.4), although transferring individual genes is relatively straightforward, it is much more difficult to transfer several genes, particularly those whose spatial position with respect to each other within the chromosome may be vital to their function.

Notwithstanding these caveats, attempts are currently being made to use genetic engineering to produce plants that withstand adverse conditions, in this case frost. Note that in our first example it is not a plant, as such, that has been engineered, but a bacterium.

▷ What advantages are there in engineering bacteria compared with engineering plants?

▶ Bacteria are more easily manipulated in the laboratory and the techniques for engineering bacteria are more advanced, giving a higher success rate. They also have a much more rapid rate of reproduction. Even if plant cells can be successfully engineered, there may be problems in regenerating whole plants from single cells (via calluses)—see Section 8.4.

Frost damage to plants amounts to many millions of pounds annually. Frost *per se* does not damage plants. What does the harm is the crystals of ice which occur inside the cells, causing structural damage and making the plants soft when they thaw out. Ice formation—the arrangement of water molecules in a precise way, instead of the more random distribution found in liquid—occurs at regularly-shaped surfaces. In practice, ice is often initially formed on the coat proteins of bacteria which live on the plant's surface. Research on one such bacterium, *Pseudomonas syringae*, resulted in the isolation of a mutant whose coat protein was altered in such a way that it no longer provided a regular surface on which ice could form. The gene involved was called '*ice*'; the mutant lacking the functional gene was called '*ice-minus*'. By specifically *removing* the *ice* gene from bacteria which live on plant surfaces, a strain called ice-minus has been developed which completely lacks the gene, and hence the protein, that allows ice crystals to form. This ice-minus strain has been used as a spray to coat susceptible crops, in the hope that they might show some frost resistance. This would only work if the ice-minus strain could successfully compete with and displace the resident surface bacteria. Whether this happens must be determined experimentally. However, the field trials of the spray have been severely curtailed in the USA by the actions of environmentalists who object to the release of laboratory organisms into the environment. Their objections revolve around the possibility of an uncontrolled spread of the test bacteria, leading to unforeseen ecological effects. In practice, it is very unlikely that the ice-minus bacteria would survive for long in the field, as mutant strains are generally less fit than the wild-type. When the trials were finally carried out it was shown that the spray did indeed afford some protection against frost; *and* the test bacteria were not detected outside the experimental plots. However, the controversy continues.

Another attempt to make frost-resistant plants, this time by engineering the plants themselves, has been made more recently. In fact the driving force behind these experiments was not to make frost-resistant crops, but to produce fruit which can be stored at sub-zero temperatures without losing its texture and flavour—a commercially important prospect. However, the 'spin-off' may prove important for food production. DNA Plant Technology, a Californian genetic engineering company, have extracted a gene from the winter flounder, an Arctic fish that can survive temperatures so low that other fish are actually frozen. The gene was copied *in vitro,* and the copies were inserted into tomato plant cells. It has been shown that the gene is expressed in the tomato fruits, which produce an 'antifreeze' protein. The protein affords resistance to freezing by binding to the water–ice interface and preventing ice crystals from forming. If the gene is effective in making tomatoes freezing-resistant, the next target for the company will be the strawberry, a notoriously poor freezer. If the gene could be incorporated into a range of crop plants it would extend not only the growing season, but also the geographical distribution of the crops. Perhaps we shall yet see wheat growing in the Scottish Highlands.

Activity 10.1 *You should spend up to 15 minutes on this activity.*

(a) What are the two main problems associated with the release into the environment of genetically engineered organisms?

(b) Do you think that the release of genetically engineered plants is likely to meet the same public resistance as did the proposed release of engineered bacteria? Summarize your view in about 150 words.

Another example of how genetic engineering might be used to make plants more environmentally adaptable is the attempt by a group at Nottingham University to make salt-tolerant rice. A wild relative of rice found in Bangladeshi mangrove swamps has microscopic structures called *salt hairs* on its leaves, in which excess sodium chloride is accumulated. At present, the group are trying to get this salt accumulation character expressed by commercial strains of rice by combining whole sets of genes from commercial and wild strains. (This is achieved by fusing protoplasts.) However, once specific gene transfer in rice becomes better established (see Section 10.3), then this may become the vehicle of choice.

10.2.2 Plant dependence on fertilizer

As noted above, modern crop plants are **heavy feeders**; that is, they require a lot of minerals from the soil. Where the soil does not contain the minerals in sufficient quantities, more must be added in the form of fertilizers. The highest-yielding varieties currently available have an absolute requirement for added chemical fertilizer—that is, they will not grow well without it on even the most fertile soils. It is no coincidence that the world's major seed companies are often large chemical companies too. By selling seeds which will need chemical fertilizer, they are creating a future market for some of their other products. Farmers who buy the seeds have no option but to purchase fertilizer as well. In the USA more than 12 million tonnes of fertilizer—about a quarter of the world's total—are used per year.

One of the major constituents of fertilizer is nitrogen.

▷ Why does a plant require nitrogen?

▶ To make amino acids and nucleotides.

The most abundant source of nitrogen is the air.

▷ What is the proportion of nitrogen in air?

▶ About four-fifths.

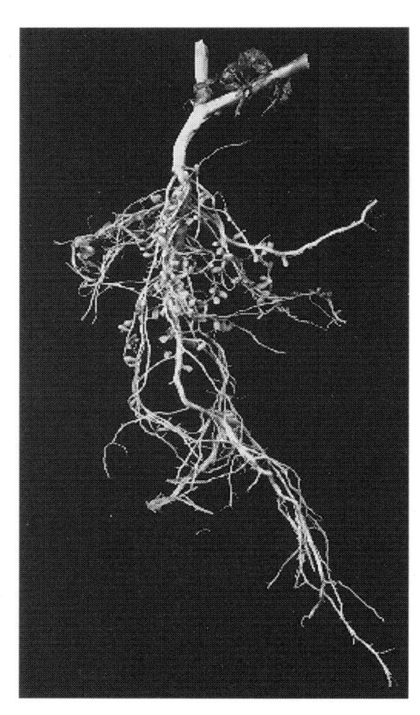

Figure 10.1 Nodules containing nitrogen-fixing bacteria on the roots of a pea plant.

Unfortunately this vast reservoir of nitrogen is not available to plants, which require their nitrogen not as the element, but as the reduced form, ammonia. In nature, nitrogen is supplied to plants mainly by soil bacteria which 'fix' the atmospheric nitrogen by enzymically combining it with hydrogen. When the bacteria die, the ammonia is released into the soil and can be taken up by plants' roots. The growth of many plants can be enhanced by the addition of chemical fertilizer. The chemical production of nitrogen fertilizer is a huge business—in 1978 the world market was estimated at $8 billion.

Some plants grow perfectly well without added nitrogen fertilizer; in particular, the legumes (peas, beans, etc.). These actually use atmospheric nitrogen which is fixed by certain bacteria living symbiotically in nodules on the roots of the plants (Figure 10.1). The bacteria receive a sheltered environment, while the plants get a supply of usable nitrogen.

There has been great interest in trying to isolate the genes responsible for **nitrogen fixation** (collectively called the ***nif* genes**) from the symbiotic bacteria. In principle, these could then be inserted into all manner of other crop species, reducing the widespread reliance on fertilizer. Predictably, however, there have been a number of problems with this approach. A major problem is that the principal enzyme involved in nitrogen fixation, *nitrogenase*, is damaged by oxygen. Therefore in a normal plant cell it would be unable to function. This is not a problem with the symbiotic bacteria because the root nodule provides a low-oxygen environment in which the enzyme is segregated. The prospect of creating such a low-oxygen environment artificially for each species of crop plant is a remote one. Currently, efforts are focused on the other side of the coin: rather than taking the *nif* genes into a hostile environment, it might be easier to alter the plants so that they too may become hosts for the nitrogen-fixing bacteria. Another problem, rather academic at present, is that if it does ever prove possible to enable plants to make their own nitrogen fertilizer, it may be found that yields actually *decrease*. This is because nitrogen fixation is an energetically expensive process: much of the energy formerly used by the plant to grow will need to be expended on fixing nitrogen—a clear case of swings and roundabouts! Although intellectually the genetic engineering of plants to fix nitrogen is a hugely interesting area, commercially it may be something of a non-starter. This is because of the competition from the chemical fertilizer industry, who stand to lose a lot of business if the demand for nitrogen fertilizer falls. The genetic engineering approach is expensive, and the relatively low cost of chemical fertilizer makes this the cheaper option, at least for the foreseeable future.

10.2.3 Resistance to pests and diseases

Up to 40% of the world's agricultural production is lost due to weed growth, pests and diseases. In particular, insects cause an enormous amount of damage to crops: it is estimated that in 1988 insects destroyed 12% of the US maize crop, and the effects of locust plagues are devastating. Pest control by conventional means is big business: the top three companies in the area, Monsanto, Dow and DuPont, together spend about $50 million per year in total on pesticide and herbicide research. Some insecticides are of biological origin, such as *pyrethrin*, which is produced by a *Chrysanthemum* species. Others are poisonous organic chemicals that are manufactured; many of these are completely synthetic and are not found naturally. As pests develop resistance to these chemicals, more and more must be used to control them. Not only is this expensive, but it results in an increasing amount of poison being deposited into the environment. One of the reasons why such synthetic insecticides are effective is that they are not natural compounds so there are not necessarily microbes that can degrade them. Until such a microbe appears and degrades the chemical, persistence in the environment may be a problem.

▷ How could a microbe able to degrade a pesticide occur?

▶ By spontaneous mutation. The microbe could, by chance, acquire the ability to exploit the pesticide as, for instance, a source of food.

We will examine protection against viruses in Section 10.2.4 and weed control in Section 10.2.5; for now we will look at pest control.

Control of insect pests is perhaps the area where genetic engineering in agriculture looks set to make the most rapid impact. In fact if you are a gardener you may already be familiar with one line of research: the use of *Bacillus thuringiensis* as protection against insect damage. *B. thuringiensis* (B.t.) is a bacterium which forms spores when growing conditions become less than ideal. When it does so, it produces protein crystals inside its cells, and these crystals happen to be toxic to many insect species, particularly to the caterpillars of Lepidoptera (butterflies and moths). Some of the most economically important plants for which the **B.t. toxin** has been used as an effective protection against insect damage are shown in Table 10.2.

Table 10.2 Some insect pests of plants against which *B. thuringiensis* toxin is effective.*

Insect pest	Plants affected
alfalfa caterpillar	alfalfa
cabbage looper	broccoli, cabbage, cauliflower, celery, lettuce, potato, melon
codling moth	apple, pear
cotton bollworm	cotton
European corn borer	maize
grape leaffolder	grape vine
green cloverworm	soya bean
Indian meal moth	stored grain
orangedog	citrus species
tomato hornworm	tomato

* There are actually several slightly different B.t. toxins each with a different range of target insects.

It is estimated that in the USA the use of B.t. toxin against the cabbage looper alone saves more than 1 000 tonnes per year of chemical insecticide. This is a significant saving, and is particularly welcome in view of the current trend away from excessive use of agricultural chemicals.

B.t. toxin has been available to gardeners for several years. It is bought as a powder and sprayed or watered on. The toxin works in an interesting way. The toxin protein is inactive in acid or neutral conditions—it is strictly speaking a *pro*-toxin—but this is cleaved and thereby activated in an alkaline environment. The guts of caterpillars are alkaline, so when plants that have been treated with the B.t. pro-toxin are eaten the pro-toxin becomes activated inside the insect. The active toxin attacks the extracellular matrix which sticks the cells together, and the gut falls apart, allowing its contents to diffuse into the insect's body. This causes a pH change that leads to general paralysis. Death occurs much later, as a result of bacterial infection.

In 1987 the gene responsible for the toxin was isolated and inserted directly into a plant, where it was shown to provide protection from insect damage. (For reasons that are not clear, it has proved necessary to engineer plants to produce the *activated* toxin rather than the pro-toxin.) This has now been achieved in several species of crop plant. Extensive field trials have been undertaken, and plants engineered with the B.t. toxin gene are indeed found to be protected against known target species. However, in cases where a crop is also affected by an insect which is not sensitive to B.t. toxin, it is likely that chemical insecticides will continue to be used.

It is reasonable to ask whether plants containing the B.t. toxin gene are safe to eat. Preliminary evidence suggests that plants containing the gene for the active B.t. toxin should be safe for consumption by humans or other animals, apart from the susceptible insects. The B.t. gene appears to be quite stable inside the plant. In fact, the whole project has been so successful, and has imposed such an environmental stress on the insects, that, unfortunately, reports are now coming in from around the world of the emergence of insects that are resistant to the toxin! Resistant insects have not arisen simply in response to the introduction of plants containing the B.t. gene; in all probability they have arisen because of the over-application, in some areas, of the toxin itself. This is the very problem that bedevils chemical pesticides: the development (by mutation) of resistant strains of pests which will then flourish until challenged by a different pesticide. This is one reason why new pesticides are continually being introduced.

In response to the appearance of pests that are resistant to B.t. toxin, genetic engineers are attempting to find other 'natural' means of pest control. One such approach to obtaining protection from insect damage (or disease) is to use the wild relatives of commercial crop varieties. Plants such as these often have remarkable properties of disease and pest resistance, and if these traits can be identified, and the genes responsible isolated, they can be incorporated specifically into commercial varieties, to improve these still further. One example of this is provided by the carrot. Carrots are very prone to attack by carrot root fly, the larvae of which burrow into the carrots. The larvae use phenolic acids, produced by the carrots, for their development. Carrot root flies are a major problem for carrot growers, because they adapt so quickly (i.e. acquire resistance) to chemical insecticides. They do not mind the cold, so can overwinter in large numbers, and there is a large reserve population in hedgerows, where they attack wild relatives of the carrot such as rough chervil and fool's parsley. Work at Horticulture Research International, at Wellesbourne, Warwickshire, has been targeted at assessing the root fly resistance of various non-commercial varieties of carrot. One species, the Libyan carrot (known to the researchers as 'Gaddafy's gift') seems to hold some promise. Carrots of this species appear to have naturally low levels of the phenolic acids needed by the root fly larvae, which therefore avoid them. It is *possible* that the low phenolic acid phenotype is due to the *lack* of a gene or genes; but it is not yet known how many genes are involved.

- Why does this matter?
- Single-gene characters present an easier target for genetic engineering than do multi-gene characters.

The B.t. toxin gene is a single gene, which is why it has been an early target of insecticide research. Another single gene that has been identified which may prove useful for pest resistance is a gene from the cowpea. This gene codes for a **protease inhibitor** —*proteases* are enzymes that break down proteins. Proteases carry out a number of vital roles in all organisms, including digestion of proteins in animal guts, and the inhibitor prevents protease activity in a wide range of insects—much wider, in fact, than the activity range of B.t. toxin. It is important, of course, not to introduce a gene whose product is detrimental to the host plant itself, as this would rather defeat the object of the exercise, but with care this can be avoided as there are several families of protease inhibitor, each of which has its own range of targets. This approach holds considerable promise for the control of pests such as the Colorado beetle, an important pest of potatoes, which is not susceptible to B.t. toxin. There are several major advantages to the use of protease inhibitors. One is their broad activity spectrum, which allows them to be used against pests which were previously only susceptible to chemical insecticides. They can be used to prevent the spread of pests resistant to B.t. toxin. They are inactivated upon cooking, and their common occurrence in human and animal foods suggests that they would not pose a toxicity problem. The major disadvantages are the high levels required to kill some insects, and possibly the need to confine gene expression to specific plant tissues (see Section 8.3.1 and Section 10.8 below).

The first example of the cowpea protease inhibitor gene being used as an insecticide was reported in 1987 by the Agricultural Genetics Company, in Cambridge. They inserted the gene into tobacco plants, and reported that high levels of the inhibitor protein provided protection against the tobacco budworm, an important pest of tobacco. The work has since been extended to the control of several pests of maize. The main drawback, as mentioned above, is that very high levels of inhibitor protein are necessary for effective insect control. Nevertheless, this shows that the procedure can be carried out successfully, and offers bright prospects for future applications.

'Let's see now – Oh yes, this must be the one with insect resistance!'

Figure 10.2 The development of insect-resistant plants. (From *Trends in Biotechnology*, June 1991, p. 199.)

10.2.4 Protection against viruses

Of course, insects are not the only pests to affect crop plants: plant viruses also cause a significant reduction in yield.

▷ How can animals be protected against viruses?

▶ Animals, including humans, can be *vaccinated*, that is, treated with a harmless part of the virus, such as the isolated coat proteins, to ensure that they will be resistant to further exposure to the virus.

Vaccination works because higher animals have a complex and sophisticated immune system (see Section 9.2.1). Plants lack such a system, so vaccination is not an option here. However, it has been found that if viral genes coding for viral coat proteins are inserted into plant DNA, *the plant becomes resistant to that virus*. Just how this works is not fully understood, but plant genetic engineers have been quick to exploit it. So far, tobacco, tomato, potato, cucumber and alfalfa have been successfully engineered to resist various viruses. One current focus of research is the African cassava mosaic virus, which can cause losses of up to 96% of the important African root crop, cassava. The major problem with viral resistance due to a single gene is that, as with insect resistance, it is likely that the virus itself will mutate, and the plant's resistance may be circumvented. But until genetic engineering can handle more complex gene systems, we must be content with the, possibly transient, resistance that we can achieve.

10.2.5 Competition from weeds

Weeds cause a significant problem to farmers. The traditional approach has been to treat whole fields with selective herbicides (weedkillers). This is relatively easy if the weed is a dicotyledonous plant in a field of a monocotyledonous crop (such as wheat; see Section 8.4), as there are herbicides that are specific for dicots. However, when the crop itself is a dicot, these weedkillers are not appropriate, although (fortunately) there are others that are. Furthermore, as with insect pests, the problem of herbicide resistance means that new weedkillers must be introduced at intervals, or else that increasingly high amounts of existing ones must be applied—sometimes so high that the crop plant itself might be damaged! Besides being expensive, over-use of weedkillers causes environmental problems due to the surface run-off and drainage of herbicides from the fields into waterways. This has given rise to the current controversy about contamination of drinking water by agricultural chemicals. If crop yields are to increase worldwide, then this is a problem that will only get worse unless a radical solution can be found.

One approach to weed control that is being actively pursued by agrochemical companies and seed merchants (often the same people; Section 10.2.2) is to insert genes for herbicide *resistance* into crop plants. This would achieve two things. First, herbicides could be applied to fields without fear of damaging the crop, no matter how much was used. Second, the seed companies could market genetically engineered seeds, in a package with the specific herbicide to which the seeds had resistance. This represents a major marketing coup for the companies involved, but is not so welcome to farmers, who would be obliged to buy the newest, 'best' seeds (indeed, these might in time be the only ones commercially available), and in addition be tied to the particular company selling the specific herbicide. This is bad enough for the relatively well-off Western farmer, but for people in the developing countries it might well mean financial ruin. Moreover, far from reducing the use of weedkillers, this approach ensures that more and more will be added to the environment. This is a good example of the two-edged sword of genetic engineering: a brilliant idea in theory, technically feasible, undoubtedly bringing benefits to some, but bringing serious problems to others. In fact, herbicide resistance genes have already been engineered into a number of crop plants, including wheat (see Section 10.4). One gene used codes for resistance to glyphosate, the active ingredient in the herbicide Roundup, made by the US agricultural chemical and (more recently) plant genetic engineering company, Monsanto. The gene has been inserted into tomato, potato and tobacco. Monsanto have come under fire from environmentalists and others for this work, but they claim that their research has focused on herbicides which need to be applied only in small amounts, have low toxicity, are rapidly degraded and stay in the soil rather than being washed into rivers. If these claims are substantiated, then the research may indeed bring environmental benefits.

We will now look at three different crops, and see how genetic engineering has, or is set to, change the way they are grown.

10.3 Rice

Rice is Asia's main staple food, and so is one of the crops whose yield must increase by around 40% to keep pace with population growth (see Section 10.1).

Research on rice has a long and well-funded history; the Rockefeller Foundation, which was a prime mover of the Green Revolution and set up the International Rice Research Institute which has developed so many of the modern commercial strains, is now interested in supporting rice genetic engineering. Although genetic engineering via protoplasts is possible in rice (Section 8.4), the efficiency tends to be low: only in about 10 cells per million will the transferred genes become stably integrated and expressed, and among these the number that can go on to form calluses, and then whole plants, is low. However, the recent application of biolistics to rice embryos has resulted in improved figures.

▷ What does 'biolistics' mean?

▶ The shooting of DNA-coated particles into cells (Section 8.4.1).

Using this technique, herbicide resistance has been transferred to rice in the laboratory. It remains to be seen how these varieties will fare in field trials (or even whether field trials will be allowed to proceed—see Section 10.2.1 above). Among the problems currently being addressed by genetic engineering are resistance to tungro virus, the improvement of rice protein, and the introduction into rice of genes for β-carotene synthesis.

10.3.1 Rice tungro virus

Rice tungro virus is unusual in that it actually consists of two viral entities. The disease-causing agent is a rod-shaped particle, but the agent that allows the virus to be transmitted from cell to cell is a separate, spherical particle. The virus is transmitted between rice plants by a number of insect species (a common mode of transmission for plant viruses), in particular the brown plant-hopper. Tungro virus causes between 40 and 60% losses in affected fields. Its effects include stunting, delayed flowering, and cancerous growth. It also causes a yellow discoloration of the grains, and this colour change means that affected rice is rejected by the consumer, even though it is safe to eat. The losses due to tungro virus are estimated at $1 billion per year. There are two avenues currently being explored by genetic engineers. One is the insertion of viral coat protein genes into the rice to make the plant resistant; the other is the insertion of the B.t. toxin gene in the hope that this might poison, and hence reduce virus transfer by, the brown plant-hopper.

10.3.2 Improvement of rice protein

The protein obtained by eating a grain of rice is actually the stored protein that is used by the rice embryo during germination. This means that it is well suited to the rice's needs, but rather less well suited to our dietary requirements. There are two major types of protein stored in rice grains. 5–10% is soluble prolamine, which is easily digested, but 70–80% is the insoluble glutelin, which is hard to digest. A group at Kyoto University in Japan is trying to put the regulatory sequences for the glutelin gene in front of the prolamine gene (Section 8.3.1).

▷ What do the regulatory sequences do?

▶ They allow the gene to be expressed in the right place (the grain), at the right time (during grain development), and to produce the right amount of protein (in the case of glutelin, 70–80% of the total).

If this hybrid DNA sequence could be constructed, it might mean that a rice grain into which this sequence has been stably integrated (and ideally with its own regulatory sequences removed to avoid possible competition between the different regulatory sequences) could make 70% of its stored protein as the useful (to us), easily digested prolamine. Of course, the effects on the rice plant itself cannot be predicted, but it would certainly improve the nutritional quality, and hence the *effective* yield, of rice for human consumption.

10.3.3 Insertion of β-carotene genes

Vitamin A is essential for vision, hence the story that carrots, a good source of vitamin A, make you see in the dark. Although this is some way from the truth, it is certainly the case that humans need a regular supply of vitamin A to maintain good eyesight. In the well-fed West, this is not a problem, but worldwide around five million people *per year* develop a painful eye condition called xerophthalmia due to vitamin A deficiency. Of these about half a million will eventually go blind. Lack of vitamin A also increases the child death rate from measles, diarrhoea and respiratory diseases. One way around the problem might be to insert genes for vitamin A synthesis into a food that many people in affected areas have access to: in this case, rice. The precursor of vitamin A is β-carotene (so called after its discovery in carrots). β-carotene is synthesized via a metabolic pathway involving a number of enzymes, each of which is coded for by a specific gene. We can refer to these as 'β-carotene genes'. Work is under way to take the β-carotene genes from cyanobacteria (a good source, and one whose DNA can easily be transferred in genetic engineering experiments) and put them in rice. Although this work is in its early stages, one potential drawback has already emerged: the engineered rice is likely to have an orange (or, actually, carroty) coloration, and as you saw above, this is likely to lead to consumer rejection. If you find the notion of consumer preferences puzzling, think how you would feel if asked to drink blue milk, for example, no matter how good it was said to be for you! This is another example of the swings and roundabouts of genetic engineering—will the benefits outweigh the perceived (aesthetic) drawbacks enough ever to allow the use of the engineered plants?

10.4 Wheat

Wheat is one of the most important crops in the world. The annual production is around 550 million tonnes, worth about $60 billion. If the yield and productivity of wheat could be increased, this would be a major step forward in feeding a growing population. Wheat has been the target for genetic engineers ever since engineering in plants was a twinkle in the breeders' eyes. However, because of technical difficulties (outlined in Section 8.4) wheat has proved very difficult to work with.

However, recent advances in tissue culture techniques (Section 7.2.2), together with the advent of biolistics, allowed a group at the University of Florida, in collaboration with Monsanto, to announce in 1991 the production of stably transformed callus lines of wheat. The genes used in the transformation were 'marker' genes, easily identifiable in the laboratory but of no commercial use, but in June 1992 the same group reported the production of fertile, transgenic wheat *plants* carrying a herbicide resistance gene. The grown plants proved to be resistant to the herbicide in the laboratory.

The importance of this work cannot be overemphasized. It is a major milestone in the history, not just of plant breeding, but of the whole modern technology of genetic engineering. The fact that it took a large number of person-years to achieve is borne out by the wording on the cover of the journal that published the result (Figure 10.3).

Figure 10.3 Front cover of the June 1992 issue of the journal *Biotechnology*.

An example of the difficulty of the project, and of the patience required by those who finally cracked it, is the fact that the transformed callus lines had to be maintained in the laboratory for 15 *months* before they would regenerate into whole plants. (However, compared with the normal generation time of wheat, a year, this is actually not too extreme. And, anyway, once produced, the engineered plants can, like normal plants, be bred by conventional techniques.) This work represents a real scientific triumph, but where will the work go from here? Given the involvement of Monsanto in the project, it is not surprising that the first gene to be transferred was a herbicide resistance gene—although it is also a useful marker, easily checked experimentally. There is no reason why any of the goals outlined in Section 10.2 should not be applied to wheat eventually. Only time will tell which will be the first genetically engineered wheat to become available commercially.

10.5 Cotton

Cotton is not primarily a food crop, but it is economically important all the same. Worldwide, an estimated 180 million people are involved in cotton production, and the annual value of the crop is more than $4 billion. Cotton is subject to attack by a very wide range of insects, costing $645 million annually. In the USA, a major producer, the intensive farming methods mean that vast amounts of pesticides are normally applied to the crops. However, with increasing concern for the environment, and the banning (due to excess toxicity) of many conventional pesticides, there is a strong case for finding alternative pest control strategies.

In the long term, there are many possible goals for genetic engineering of cotton (see Figure 10.4). Some of these, such as the complex characters of salt or drought tolerance, are likely to be very long-term indeed.

▷ Why should this be?

▶ These complex characters are likely to be controlled by many genes, and are therefore not so readily amenable to genetic engineering (Section 7.4).

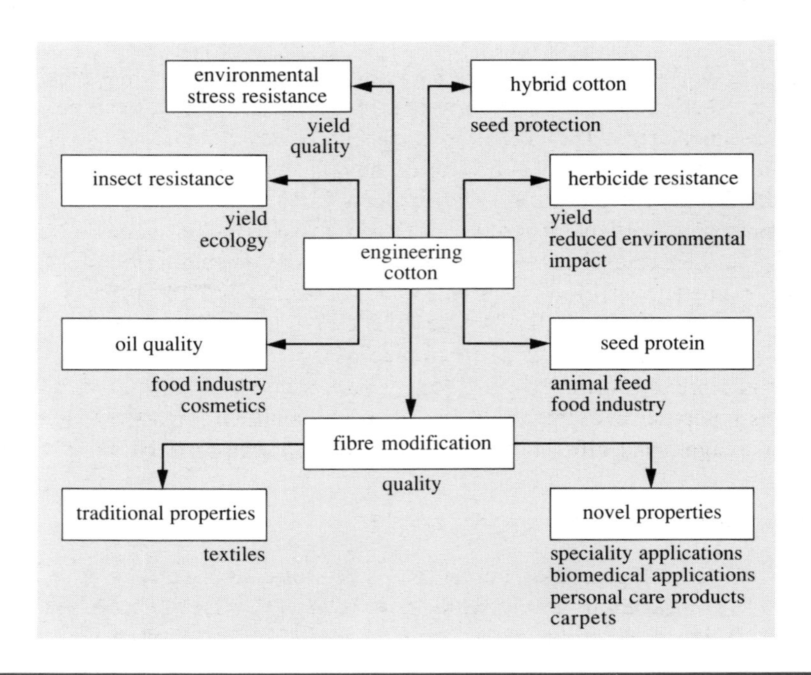

Figure 10.4 Genetic engineering goals for cotton. Engineering cotton to withstand environmental stresses such as heat, cold and drought might increase quality and yield, as could the introduction of resistance to insect pests and resistance to herbicides. These latter improvements would also reduce the adverse environmental effects of chemical application. Increasing the quality of the seed protein, and of the oil it contains, could increase the value of the seed for the animal feed and food industries. Improvement to fibre quality would benefit the textile industry, e.g. better heat absorption and retention could make cotton better suited for winter clothing. Novel properties, such as increased liquid absorption, would be useful in the personal care sector (e.g. for cotton wool).

The relative simplicity of transferring single-gene characters into plants means that other targets are likely to be achieved in a shorter time frame—indeed, some have already been achieved (see below). Cotton has been successfully engineered in the laboratory using both the T_i plasmid method and biolistics (see Sections 8.3 and 8.4.1).

▷ What is the main difference between these techniques?

▶ The T_i plasmid method involves transferring DNA by means of a *vector*; biolistics involves shooting *DNA-coated particles* directly into cells.

10.5.1 Resistance to pests

Given the vulnerability of cotton to insect pests, the production of genetically engineered insect-resistant plants is an obvious goal. In 1989 the B.t. toxin gene was inserted into cotton cells using a T_i plasmid vector, and whole plants grown from the engineered cells. The B.t. toxin was made by the cotton in large quantities (up to 0.1% of the total cell protein), and the recombinant plants showed significant caterpillar resistance in field trials.

Because cotton fibres are not eaten by humans, safety considerations with respect to toxicity need be much less stringent, and the use of strains containing the B.t. toxin gene is expected to reduce by 50% the amount of pesticides used against Lepidoptera (butterflies and moths; Table 10.2). Protease inhibitor genes may also play a useful role in cotton, as may toxin genes from spiders (whose toxins paralyse flies).

Cotton is damaged by weeds, especially bindweed and morning glory, which are a particular nuisance around harvest time. Currently, control is by conventional herbicides, most of which work by inhibiting a particular cellular enzyme. But unfortunately these damage the cotton plants too.

▷ Given this information, suggest ways in which cotton plants could become resistant to herbicides.

▶ The plant might overproduce the target enzyme, or a mutation might arise in the gene coding for the enzyme such that it was no longer affected by the herbicide.

A third way is that the plant might acquire a new gene to destroy the herbicide. All these strategies have been used by cotton breeders so that herbicides can be used to kill the competing weeds. The bacterium *Klebsiella ozaenae* contains a gene which codes for nitrilase, an enzyme that detoxifies the active ingredient of the herbicide bromoxynil. This gene has been inserted into cotton to produce bromoxynil-tolerant lines. Bromoxynil is particularly useful against bindweed, so this work will allow the herbicide to be used extensively on cotton fields without damaging the cotton crop. Work is in progress to develop cotton that contains a gene conferring resistance to the herbicide glyphosate (Section 10.2.5). This herbicide inhibits 5-enolpyruvyl-shikimic acid 3-phosphate synthase, an enzyme involved in the synthesis of some amino acids. Glyphosate is a broad-spectrum weedkiller, and has a relatively low environmental impact, so is a popular one. If glyphosate resistance could be imparted to cotton, this might make a significant difference to the cost of losses due to weed competition.

10.5.2 Improvement to cotton fibres

As you can see from Figure 10.4, among the targets of cotton genetic engineering some relate not to the ability of cotton to grow better, but rather to other features, such as the quality of the cotton fibres obtained from the plant. Cotton fibres are 87%

cellulose. The enzymes involved in cellulose synthesis have not been identified in cotton, although they have been studied in bacterial systems. It is likely that the non-cellulose components of cotton, such as pectins and proteins, are important in determining fibre properties, such as strength. A cotton fibre develops as the result of the expression of thousands of active genes, and those genes whose products are *specific* to the fibres themselves can, in principle, be identified.

Once more is known about the molecular basis of cotton production, attempts can then be made to 'improve' the genes in one way or another, perhaps in the manner of rice protein improvement outlined in Section 10.3.2 above, or even to introduce completely new properties to the fibres.

Activity 10.2 *You should spend up to 30 minutes on this activity.*

Broadly outline the techniques that might be used to introduce a new gene into cotton to give the fibres an improved capacity to absorb water. Assume that the improved water absorption property is due to the presence in the fibre of a single protein.

You will probably need to refer to earlier sections, particularly Sections 5.3, 6.4, 8.3 and 8.4.

It is expected that cotton will be among the first genetically engineered plants to reach the market. As cotton fibres are not eaten, it is expected that the safety and efficacy testing required for any improved product will be relatively straightforward (see Figure 10.5). Note that the cotton seed protein and oil, which *are* eaten (Figure 10.4), are both purified products, so should not contain any B.t. toxin; thus there would be no question at all of any toxicity risk.

'Looks promising Wilkins – that's fine for shrink- and wrinkle-resistance – now for the acid test.'

Figure 10.5 Genes for jeans. (From *Trends in Biotechnology*, May 1992, p. 168.)

10.6 Targets in animal breeding

By comparison with genetic engineering in plants, work with animals is in some ways much less advanced. In part this is due to the ethical problems associated with work on animals, particularly mammals (see Sections 8.7 and 10.9), and in part it is due to the more difficult technical aspects.

▷ Which technique has simplified the problem of getting DNA into target plant cells?

▶ Biolistics (Section 8.4.1).

Unfortunately, although biolistics may be a powerful method for inserting genes into plant cells, the technique is not yet highly efficient in animal cells. The reasons for this are not fully understood, but are thought to be due to the extreme fragility of animal cells, particularly the comparatively huge egg and embryonic cells which are used for these manipulations (Section 8.5.1 and Section 10.6.1 below). Whatever the reasons, until the biolistics approach becomes more efficient, the only 'reliable' method of genetic engineering in animals is by injection of DNA into cell nuclei. This is obviously extremely labour-intensive, and has a very low success rate. Even in laboratory mice, the best studied species, often less than 10% of treated eggs result in a genetically modified animal that can pass on the engineered genes to its offspring. However, once a transgenic animal *has* been produced, it can readily be reproduced by modern breeding methods. If the animal is a male, sperm retrieval and storage methods, followed by artificial insemination, can ensure a large number of offspring carrying the engineered gene. Even if the animal is female, egg retrieval and subsequent embryo splitting (see overleaf) can mean a reasonable number of offspring from the original animal. Alternatively, hormone treatment can be used to induce the simultaneous ovulation of as many as 30 eggs (**superovulation**) from one cow (instead of the usual one or two). These eggs can be removed non-surgically from the cow, fertilized *in vitro*, have foreign genes inserted into them if required, and the resultant embryos can then be implanted into hormonally-primed surrogate mothers. The surrogate cows will subsequently give birth to élite calves (see Figure 10.6).

Figure 10.6 A group of 19 calves together with their genetic mother. These calves were derived by superovulation of the donor cow, followed by artificial insemination and recovery of the resultant embryos, which were transferred to hormonally-primed surrogate recipients. In this instance, 28 embryos were recovered and resulted in the birth of 19 calves. (*Note*: in this case, fertilization occurred *in vivo*.)

10.6.1 Embryo splitting

All mammalian embryos follow a similar developmental path on their way to becoming newborn animals. The fertilized egg, which is a very large single cell, undergoes a series of divisions (without net growth) whose result is to produce a **blastocyst** containing two cell populations (Figure 10.7). One of these populations (the *trophectoderm cells*) contributes to the placenta; the other (the *inner cell mass*) constitutes the 'real' embryo, i.e. the cells which will divide and differentiate to form a whole new animal.

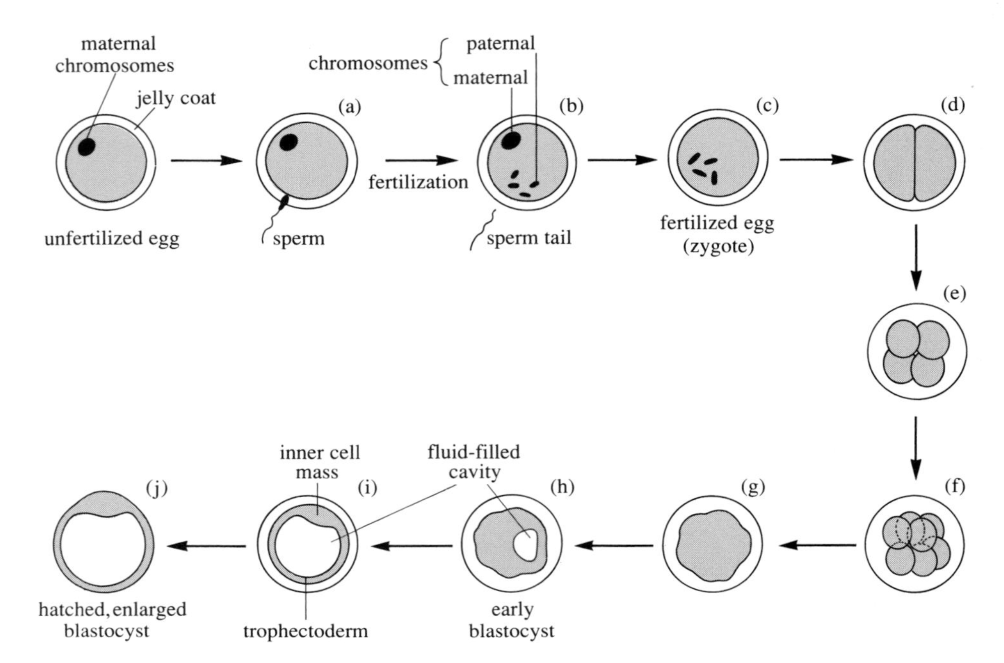

Figure 10.7 Stages in early mammalian development. (a) A sperm penetrates the jelly coat and fertilization occurs. (b) The paternal chromosomes from the sperm are released into the egg cytoplasm. (c) In the fertilized egg, maternal and paternal chromosomes come together, allowing the first mitotic division to take place (d). Successive divisions follow (stages (e) and (f)), resulting in (g) a ball of 16 to 32 tightly packed cells. (h) A fluid-filled cavity appears within the ball of cells, and grows until the cells are stretched around the outside of the ball. This structure is called the *blastocyst*. It consists of two cell types, the *trophectoderm* and the *inner cell mass* (i). The blastocyst continues to expand until (j) it 'hatches' out of its jelly coat. It is then ready to implant in the wall of the uterus. The trophectoderm cells contribute to the placenta, while the inner cell mass gives rise to the whole embryo.

In the early stages, up to about the eight-cell stage (though this varies with species), each of the embryo's constituent cells—even just the nucleus from each cell—is **totipotent**; that is, it can form any or all of the animal's tissues given the correct cytoplasmic signals. This means that if the cells are separated artificially, each can give rise to a complete embryo (Figure 10.8a). The figure also shows that *pairs* of cells separated at the eight-cell stage may regenerate an embryo. The embryos thus obtained can be transferred to surrogate mothers and grown to term (Figure 10.8b).

This process happens naturally when identical twins, triplets or quads arise. It means that one fertilization event can give rise to several identical offspring, to produce a clone. Any genetic engineering carried out on the original embryo can therefore be amplified to give a clone of transgenic animals.

▷ What is the main difference between this kind of clone and a clone of bacteria?

▶ The number of members. A bacterial clone may have millions of identical members, but a clone of animals will have only a few (twins, triplets, etc.).

What, then, are the targets for genetic engineering in animals? So far efforts have been directed towards farm animals, though no doubt in time the companion animal (i.e. pet) sector may also want to benefit from a technique that allows a choice in characteristics—to produce the clawless cat, for instance. There have been two main thrusts to research in this area: making farm animals more productive in terms of meat, milk, etc., and getting the animals to produce novel substances such as drugs. This latter process is colloquially known as *pharming*. We will look at examples of each of these.

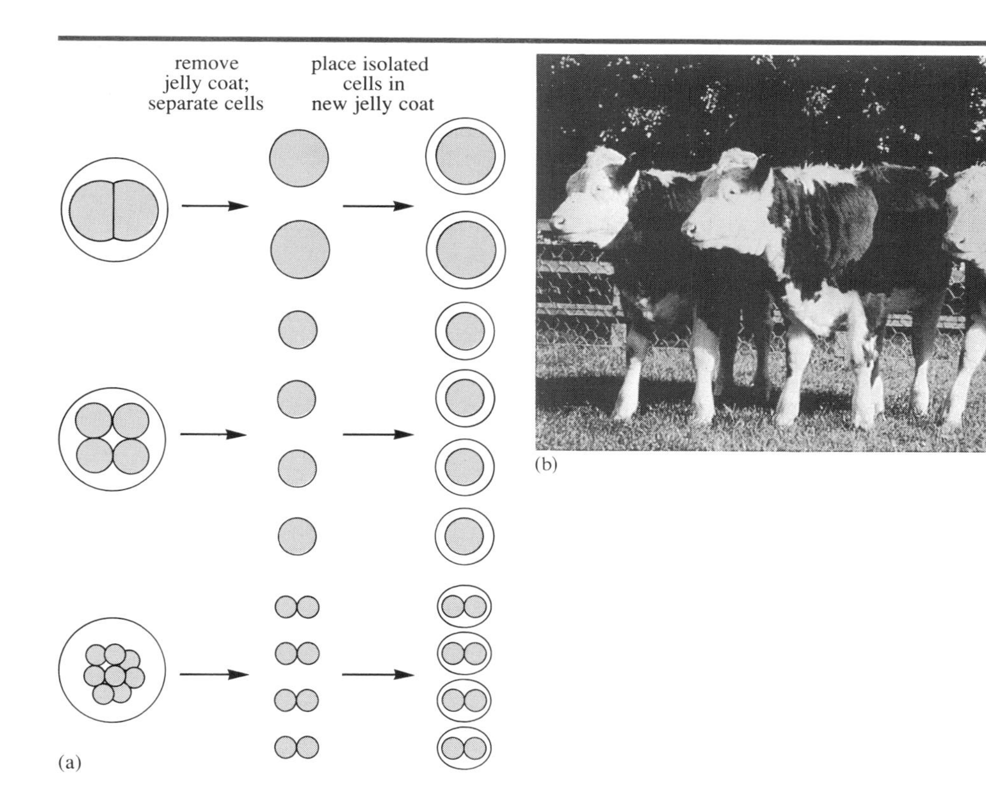

Figure 10.8 (a) Separating embryonic cells for cloning animals. (b) Genetically identical triplet calves. These calves were produced by separating the cells of an early embryo into quarters and transferring the resultant embryos to recipient surrogate mothers. (One of the four embryos failed to survive.)

10.7 Improving farm animals

The standards of food production achieved by conventional animal breeding techniques are often quite astonishing. A modern dairy cow, for instance, can produce 10 000 litres of milk each year. However, this level of production relies on intensive farming methods, particularly a very rich diet. This is clearly out of the reach of many subsistence farmers in the developing world, whose cattle often have to graze very poor pasture, and can produce only 500 litres of milk per year. There seems to be an opportunity here for introducing genes into cattle to enable them to sustain high milk yields on a poor diet.

▷ Can you think of another way to overcome the problem?

▶ Engineer the *plants on which the cows graze* so that they contain more nourishment.

The Indian Department of Biotechnology has a cattle herd improvement programme which aims to double annual milk production from 40 million to 80 million tonnes in 10 years. One way of doing this is to insert copies of the *bovine growth hormone* (**bovine somatotropin, BST**) gene into separated embryonic cells of dairy cows. The product of this gene, besides increasing growth rate, muscle content and overall size in the adult animal, also improves milk yields by as much as 25%. India has more than 250 million cattle and 70 million buffaloes, most of which are poor yielders. They will be used as surrogate mothers to produce the new, élite, calves—thereby upgrading the herd. With this major focus on animal engineering, progress is likely to be substantial.

Wool production

Meat and milk are not, of course, the only products obtained from farm animals. Wool is another major product, and in the late 1980s annual worldwide production was around two million tonnes. Australia produces one-third of the world's wool, and the main sheep breeds are high-yielding, fine-wool sheep, based on the Merino. Selective breeding has produced sheep with high fleece densities: there are 100 million hair follicles per sheep, and wool production is limited by competition between the follicles for the available nutrients (brought to the skin by the blood). Although each breed of sheep has a genetically determined maximum yield of wool, individual sheep can vary fourfold in wool production, depending upon the amino acid supply.

▷ Why are amino acids so important?

▶ Wool is a protein, called keratin, and is assembled from amino acids in the blood supply.

It turns out that the most important, limiting, amino acid is cysteine. This has an —SH group in it, which can react with other similar groups to form disulphide bridges (see Box 5.1 and Section 9.3.2). Disulphide bridges link peptide chains together in a particular orientation, and contribute substantially to the higher-order stucture of many proteins.

Sheep are ruminants; they have four stomachs, in one of which, the rumen, symbiotic bacteria carry out much of the digestive process, particularly of cellulose-containing vegetable matter, which mammals themselves cannot break down. The symbiotic bacteria digest the proteins eaten by the sheep, and use the resulting amino acids to synthesize other proteins which are used for their *own* growth, and which, in the fullness of time and further down the digestive tract, the *sheep* will digest and use. The proteins synthesized by the symbiotic bacteria contain less cysteine than the proteins eaten, i.e. those contained in the grass. When the bacteria synthesize their own proteins, they excrete the excess amino acids, which they do not require, back out into the rumen. Because sheep cannot absorb amino acids from the rumen, the free amino acids, in particular a lot of cysteine, are lost. (There are other bacteria in the rumen that take up cysteine and excrete the sulphur component as hydrogen sulphide gas, which is burped out.) What the sheep *finally* absorbs (the amino acids contained in the proteins of the symbiotic bacteria) has an insufficient cysteine content for the growth of high-quality wool. This is why cysteine is the limiting amino acid for commercial wool production.

In fact, matters are slightly more complicated than this, but the very complexity of the system offers two options for genetic engineering: to engineer the sheep, or to engineer the symbiotic bacteria. Both these approaches will be considered.

▷ What would be the aim of engineering (a) the sheep, and (b) the bacteria?

▶ (a) The sheep could be made to use the excess cysteine that is released in the rumen. (b) The bacteria could be made to produce proteins containing more cysteine, thereby preventing its loss to the sheep.

Remember that the sulphur component of cysteine is lost as hydrogen sulphide in the rumen. One way to overcome this loss would be to transfer the genes for the enzymes involved in the resynthesis of cysteine from sulphide, into the cells of the sheep's rumen. Only two enzymes are involved in the reaction, so it presents a reasonably good target for genetic engineering. Suitable genes have been isolated from a bacterium, *Salmonella typhimurium*, and after attachment to appropriate regulatory sequences, the genes have been transferred to sheep, where they were shown to be expressed.

▷ What does 'appropriate' mean in this context?

▶ Regulatory sequences that can be recognized by sheep rumen cells—remember that bacteria and mammals may recognize *different* signals.

Ways have not yet been found of limiting expression to the rumen, but this is not likely to be a problem for the sheep, since the genes are only active where there are high concentrations of sulphide. High concentrations occur only in the rumen. The success rate for producing engineered sheep is low, ranging from 0.1 to 2%. Of course, other sheep can be bred from these successes, but it is likely to be some time before large flocks of transgenic sheep are roaming the outback.

The option of engineering the symbiotic bacteria from the rumen is, paradoxically, much more difficult. This is because in order to make the bacteria produce proteins containing more cysteine, a very large part of their metabolism would have to be changed, and as we have repeatedly pointed out, this presents a fearsome task for genetic engineering at present. In any case, this might make the bacteria less able to survive. For once, it seems that, for the purposes of genetic engineering, higher organisms have the edge over bacteria!

Livestock disease

We cannot leave this section without mentioning one final target for improving farm animals, even though research in this area is in its infancy. This is the fight against livestock disease. Disease, or the threat of it, is a cause of huge losses to agriculture. In the industrialized world, losses run at about 17%, but in the developing countries this can rise to a massive 35%. The top two animal diseases, in terms of value lost, are foot and mouth disease, and mastitis, and vaccines against these are under development.

Although much disease can be prevented by administering suitable vaccines (Section 9.3.1), another possible approach is to make the animals inherently resistant to the diseases, as with plants. One fatal disease, trypanosomiasis, has severely limited the use of cattle for draught, as well as food, purposes in Africa, because the tsetse fly, the transmitter of the disease, is so widespread—the disease is endemic in 36 sub-Saharan countries. Resistant cattle would be a boon not only from the food production view, but from the agricultural labourer's standpoint too. Work is already in progress to transfer the gene for resistance to trypanosomiasis from wild, resistant, livestock to domesticated varieties.

10.8 Pharming

The idea of producing high-value goods cheaply is obviously an attractive one. As explained in Chapter 9, the pharmaceutical industry has been quick to exploit the features of genetic engineering that allow it to make easily and sell large quantities of materials that traditionally have been obtainable in only small quantities. **Pharming** is the production in farm animals (and to some extent plants) of pharmaceuticals. This is not so much a question of 'improving' the organism, but is more one of making the organism the equivalent of a microbial 'factory' (see Chapter 3). Pharming is being actively researched by many companies. There are four main reasons for this:

1 There exists the possibility of very high volume productivity. For example, a dairy cow can produce 10 000 litres of milk per year. If a pharmaceutical could be produced in the milk at, say, $35\,g\,l^{-1}$, then 350 kg might be produced per animal per year. This represents a lot of pharmaceutical in any terms.

2 Low operating costs. Once the producer animal is made, it can be kept under normal conditions for that animal—this is much cheaper than commissioning a new production plant.

3 Animal or plant cells provide the ideal environment for the production of eukaryotic proteins. All the regulatory and post-transcriptional and post-translational

modification systems are already present in the cells (see Box 5.2 and Section 7.2.1), although in fact some of the sugar side-chains added may differ between species.

4 In theory, there would be unlimited reproduction of the factory!

The main thrust of the research has been to get the 'pharmaceutical genes' (just what these are we shall see shortly) expressed in a part of the organism that is not used for anything else (by humans, at any rate). For example, in a potato plant, the only 'useful' part of the plant is the potato itself. All the stem and leaves are discarded. If they could be made to produce a pharmaceutical, then these parts too could be harvested at little extra cost, and the pharmaceutical purified from, say, the leaves. It would be vitally important, of course, to make certain that the pharmaceutical was produced *only* in the leaves—nobody wants to eat potatoes that contain drugs (although, commercially, it is more likely that such engineered potato plants would be used exclusively for drug production). In fact, a potato of this sort has been constructed, producing not a pharmaceutical, but a plastic called polyhydroxybutyrate, PHB. The required gene was obtained from a microbe, *Alcaligenes eutrophus*. The potato leaf cells can produce up to 80% of their cell contents as PHB, which makes the plants difficult to harvest! The same principles apply to animal pharming. In an animal destined for the meat market, it is important that no foreign substance should be present in the animal's muscles. However, if the substance could be made only in the milk, then this milk could be used as the starting material for extraction and purification of the drug. Milk is a very good tissue for this type of use. It is made automatically for a large part of the year (9 months in dairy cows), the infrastructure for collecting and delivering it already exists, and, in the Western world at least, there is a milk surplus, and much is currently thrown away, causing considerable pollution. One gene which is expressed specifically in milk is that for casein, a major milk protein, and indeed the casein regulatory sequences have been used in much of this work.

The first demonstration that pharming would work came in 1987 when transgenic mice were produced that would secrete human t-PA (an enzyme that helps to dissolve blood clots; Sections 7.2.2 and 9.3.2) in their milk. Since then, production of foreign proteins has been obtained in the milk of sheep, goats and cows. Perhaps the most striking example of this is the work of a group in Scotland. They have produced transgenic sheep which produce human α-1-antitrypsin in their milk. α-1-antitrypsin is a serum protein which prevents emphysema of the lungs. It has been approved in the USA as replacement therapy for people who are genetically deficient in its production, and who are at risk from emphysema. Currently, the protein is purified from human blood plasma where it is present at a concentration of $2\,\text{g}\,\text{l}^{-1}$. However, the demand for the substance is so high that an alternative source would be welcomed. In this work, the human α-1-antitrypsin gene was put under the control, not of the casein regulatory sequences, but of those for sheep β-lactoglobulin, another milk protein. Eggs were collected, fertilized *in vitro*, injected with the recombinant DNA, and the resultant embryos implanted into surrogate mothers (see Section 10.6). Of 152 embryos used, four females and one male had incorporated copies of the gene into their own chromosomes. All the animals developed normally, and the females later gave birth to a mixture of transgenic and non-transgenic lambs.

▷ Why was there a mixture of offspring?

▶ The transferred genes were presumably only stably incorporated into *some* of the chromosomes. So after meiosis, some of the gametes would contain the inserted genes, giving rise to transgenic offspring, while the others would not.

The milk of all the engineered females contained biologically active α-1-antitrypsin, indistinguishable from the 'normal' human sort, and one produced it in quantities of up to $35\,\text{g}\,\text{l}^{-1}$. If these levels could be consistently reached, then one moderately sized herd could produce several thousand kilograms during one lactation season. This, of course, raises ethical issues. It may seem like one more example of human exploitation of farm animals. But the motive behind it is to cure a serious human disease. So, once again, this involves a tricky moral balancing act, as did the OncoMouse work mentioned in Activity 8.1. Another factor to consider is that the engineered animals are valuable, and will have a life of relative luxury. The ethics of this are inevitably complex, and, as you will discover if you discuss this, the matter can promote heated argument. Which side is right? We leave it to you to decide.

Early in 1992, in the laboratory, mice were successfully engineered to produce the human protein which is deficient in cystic fibrosis. As discussed in Section 9.4, cystic fibrosis is the disease caused by an abnormal ion transport system across cell membranes. The cystic fibrosis protein (CFTR) is a membrane protein, and when placed under the control of the casein regulatory sequences in the transgenic mice it is found in milk fat globules, apparently in the membranes enclosing the globules. This research, by a US group, will allow enough of the protein to be produced for the precise molecular defect to be identified, and, with luck, better therapeutic strategies adopted for this disease.

10.9 The BST controversy

Bovine somatotropin (BST) is a growth hormone which is essential in cows for growth, muscle development and milk production (among other things). Very early on in genetic engineering research on animals, it was shown that extra copies of growth hormone genes could increase size beyond the norm (see Section 8.5.1). As you saw in Section 10.7, putting extra copies of the BST gene into cows can increase milk production substantially. More than a decade ago, tests began on BST made in bacteria via rDNA technology to assess its efficacy and safety for widespread use in the dairy industry. These experiments involved injecting the BST *protein* into cows, rather than the BST gene. Since then, BST has been tested on more than 21 000 cows worldwide. Yet there remains considerable opposition, both in the USA and in the EC, to the use of BST injections for enhancing milk production. There has been a boycott in the USA by some dairy farmers, four major supermarket chains, and a large number of consumers. The dairy farmers' worries include the fact that the increase in production will depress prices, and force smaller operators out of business. Other worries centre on two questions: is the milk from treated cows safe for human consumption, and is it ethically justifiable to obtain milk in this way? We ask you to consider these questions in Activity 10.3.

Activity 10.3 *You should spend up to 1 hour on this activity.*

Extracts 10.1–10.3 refer to the use on cattle of BST (produced by techniques of genetic engineering). The scientific and political background to the controversy is described and different points of view put forward.

(a) Read Extracts 10.1 and 10.2. Then write a brief article on BST (about 200 words), for newspaper publication, using these extracts as your source of information. We suggest you approach this task by looking at each paragraph of the extracts in turn, and pick out the main ideas, making notes where necessary. You may find it helpful to underline or highlight key ideas or terminology. Then put your selected points together, in a concise and logical sequence that tells the story. You will probably need first to produce a draft, which is then refined and shortened. (*Note*: in these two extracts BST is written as bST.)

(b) Now read Extract 10.3. Using this, and Extracts 10.1 and 10.2, list as a series of numbered points, the arguments (i) for and (ii) against the use of BST in cattle.

Extract 10.1 From *Biotechnology*, February 1992, p. 147.

Putting the bST human-health controversy to rest

by Henry I. Miller

No matter what else it elicits, bovine somatotropin (bST) has stimulated one of the most vigorous but misdirected and gratuitous controversies ever to have accompanied the testing of a production drug in agriculture. bST is hardly a scientific novelty. That it can increase milk production by up to 25 per cent has been known for more than half a century, while the biochemistry and physiology of recombinant DNA-derived bSTs have been studied exhaustively for a decade. bST has been tested on 21 000 cows worldwide and is the subject of more than 900 research papers. According to a report of six New England commissioners of agriculture, the hormone 'has been researched more than any other new technology'.

Yet the use of bST has come under strong attack, specifically by certain consumer groups and dairy farmers who oppose the marketing of milk from cattle in experimental herds. Activists have launched a boycott of dairy products from treated herds, even though consumption of the products was authorized by the FDA (Food and Drug Administration, Bethesda, MD) more than six years ago. (The FDA allows the commercial marketing of milk and meat from animals treated with experimental drugs—but only after scientific studies show that the food is safe for human consumption.)

The boycott nevertheless has been joined by four major supermarket chains, two of America's largest manufacturers of dairy products, and a well-known ice cream producer in Vermont. In addition, Associated Milk Producers Inc. (San Antonio, TX), the country's largest dairy cooperative, announced that its 21 000 members will not give recombinant DNA-derived hormone to their cows. Most recently, the state legislatures of Wisconsin and Minnesota decided to ban the use of bST temporarily, pending its final review by FDA.

Typically, two charges are levelled at bST—that bST-stimulated milk is unsafe for consumption, and that milk production increases afforded by bST use will lower milk prices. Only the first falls within the purview of FDA. As with all new veterinary drugs, the Agency's review of bST is limited by law to the hormone's safety for humans, treated animals and the environment, and to its efficacy. The second concern, like all others for economic consequences, is best judged in the marketplace by consumers of agricultural technology (dairy farmers) and of agricultural products (grocery patrons).

Much of the concern about the safety of milk from bST-treated cows appears to stem from the source of the drug: new biotechnology. All the companies that have applied for FDA approval to market bST produce it through recombinant DNA technology. FDA is experienced in this area, having already approved more than 500 products of new biotechnology and more than 1 000 clinical trials of human drugs and biologics with these products. As a science-based regulatory agency, FDA is committed to evaluate biotechnology products by the same high standards of product safety that apply to all similar regulated substances. The Agency will soon complete its final evaluation of the safety of bST to the target animals, and of its efficacy.

Science knows bST

That milk from bST-treated cows is safe for humans was amply established well before FDA authorized the marketing of milk and meat from experimental herds. As early as the 1950s, bovine growth hormone was shown to be inactive in humans, when physicians tested it as a remedy for dwarfism. It has been also well-documented that the recombinant bSTs are minimally, if at all, different in amino acid structure or physiological functions from the naturally occurring hormone, and that any small differences do not affect the drug's inaction in humans.

The FDA remains convinced that, on the key question of human safety, products from bST-treated herds pose no risk. If it had entertained any doubts, FDA would not have allowed products from experimental herds to be marketed.

Nevertheless, to answer the persistent few vocal individuals who doubt bST's safety, in 1990 FDA took the unprecedented step of submitting to extensive peer-review evaluation the pivotal scientific evidence about the effect of bovine growth hormone on human health. The authors—Judith C. Juskevich, an FDA consultant, toxicologist and pharmacologist, and C. Greg Guyer, a chemist in FDA's Center for Veterinary Medicine—summarized and analysed 69 scientific studies, some of which include previously undisclosed proprietary data from bST manufacturers. The study (*Science*, **249** (4972), 24 August 1990) concluded:

(a) bST is naturally present in milk of all cows; in cattle treated with bST, no more reaches the milk than the upper limits of normal bST levels. When taken orally, bST is broken down into inactive fragments during digestion and has no effect on human health. Moreover, 90 per cent of bST in milk is destroyed by pasteurization.

(b) bST is biologically inactive in humans even if injected, because its amino acid sequence is about 35 per cent different from human somatotropin. The bovine growth hormone is 'species-specific' (i.e. it does not trigger responses in higher species, such as humans and monkeys).

(c) The results of bST toxicity studies were negative, even when rats were fed for 14 days the equivalent of daily doses 100 times higher than those administered to dairy cattle.

In the spirit of openness, and because some controversy remained, the Agency followed the publication of the article by another unprecedented step: it requested that the National Institutes of Health (Bethesda, MD) convene a special panel of experts to evaluate all human health aspects of bST, including the safety of meat and dairy products from treated herds. At the end of its meeting in December 1990, this panel concluded that 'the composition and nutritional value of milk from rbST*-treated cows is essentially the same as milk from untreated cows', and that 'meat and milk from treated cows are as safe as those from untreated cows'.

These conclusions, reflecting virtual unanimity in the scientific community, ought to put the controversy to rest. FDA will continue to evaluate whether bST is safe for cows and the environment and whether it increases milk production as claimed. If any of these questions is answered in the negative, the drug will not be allowed on the US market. On all of these questions, FDA is confident that any regulatory decision based on sound science will bear the full weight of public scrutiny and, most important, that public health will continue to be protected.

* rbST = recombinant bST, i.e. the bST protein produced by genetic engineering.

Extract 10.2 From *Biotechnology*, February 1992, pp. 148–149.

bST & the EEC: Politics vs. Science

by William Vandaele

Since July 1987, it has been mandatory in the European Economic Community (EEC) to obtain a favourable opinion from the Committee for Veterinary Medicinal Products (CVMP, Brussels, Belgium) before a member state (national) authority can approve a medicine produced by biotechnology. The CVMP is a European Commission (EC) body composed of senior regulatory experts from each of the twelve EEC member states. The intention was that having a centralized opinion from this committee would accelerate the national regulatory processes so that the EEC can compete with the US and Japan.

In July 1987, Monsanto Europe (Brussels) introduced its dossier for bST (bovine somatotropin), the first recombinant animal health product intended to modulate the milk production. A number of delays ensued, partly because the system was new, and partly because some countries had fundamental reservations about the use of productivity enhancers. It was not until March 1991—nearly 4 years later—that the CVMP issued a favourable opinion: the Monsanto bST product, according to CVMP, satisfied the criteria for quality, efficacy and safety. Most significantly, the CVMP unanimously agreed that bST posed no risk to consumers of meat or milk from supplemented cows, and that the meat, milk and processed milk products were unaffected by bST administration to the cows. Despite this, some member states requested confirmatory data on some aspects related to cow safety, data which Monsanto has subsequently provided.

In parallel with the EC regulatory process for biotechnological medicines, the Commission in early 1991 issued a position paper on competitiveness in biotechnology. Its goal was to encourage biotechnology and to prevent Directorates General (DG, such as DG VI, which is responsible for agriculture) from issuing regulations which contradicted overall EC policy. To judge from what has happened with bST, that goal is still distant.

DG VI's independent moratoria

DG VI has twice (in 1989 and 1990) proposed twelve-month moratoria during which member states could not authorize the administration of bST. These 'evaluation periods' were ostensibly for DG VI to obtain the results of additional studies it had commissioned on the potential social and economic impact of bST on the Common Agriculture Policy (CAP). DG VI's twin goals for the CAP are (a) to minimize surpluses (i.e. to minimize the cost of purchase, storage and disposal of excess production) and (b) to maximize farm employment. Dairy quotas, which address both these issues, are already in place, but despite that, DG VI market managers believe that any productivity or efficiency enhancing product is counter-indicated.

Towards the end of 1991, with the previous bST moratorium coming to an end, the Council of Agricultural Ministers had to decide on its subsequent action on bST. The product approval process was close to its end for Monsanto's bST product; a CVMP opinion on Elanco's product is also imminent. There were, therefore, no further side-alleys down which legislative arguments could be diverted. What would the decision be?

To the surprise and disappointment of a lot of European thinkers, DG VI has proposed a further two year extension of the moratorium, a measure the Council of Ministers is likely to 'nod' through. The official report justifying this new delay has not yet been published, but DG VI's views are already known. With the CAP due for reform (stimulated by GATT discussions), the bureaucrats consider that the imposition of severe dairy quotas (a very likely measure) would be inconsistent with the authorization of a productivity enhancer such as bST. They hold this view even though companies have provided data showing that under practical farming conditions in Europe, the same quantity of milk can be produced from bST-stimulated herds—but at a lower cost.

The other argument advanced by opponents of bST within the Commission is that a possible negative consumer reaction against bST might reduce milk consumption (and hence dam the milk lake). This is a theoretical risk. Against it must be set the results of a survey conducted in UK after a major (and largely negative) press campaign publicizing the fact that milk from bST trial sites was reaching the market. This survey showed that, despite the press coverage, consumers had not changed their dairy product purchasing habits.

Biased opinion polls?

DG VI sponsored its own 'studies' to assess consumer response to bST. Its questions, which in effect asked, 'Do you want to drink milk with hormones?' unsurprisingly produced what we at FEDESA would regard as a biased result. The question, as posed, is, in any case irrelevant: there is scientific unanimity that the composition of the milk—including levels of 32 'hormones' normally present in milk—from cows supplemented with bST and milk from untreated cows does not differ significantly.

In reaching its proposal for a further two-year moratorium, DG VI did consult independent experts—the animal welfare section of the Scientific Veterinary Committee (SVC), an advisory body constituted by the DG VI. The SVC concluded that there was insufficient *published* information to conclude that there were no potential animal welfare concerns with cows administered bST. The consultation with SVC seems inappropriate on two counts. Firstly, the SVC, since it is not a regulatory body, did not have access (nor did it request access) to the copious unpublished studies on animal welfare specifically requested by (and seen by) the CVMP. After it had evaluated the complete data (which included studies in more than 10 000 cows), CVMP concluded that 'Monsanto's bST does not present any undue risk for the health or welfare of the treated cows'. Secondly, it is ironic that DG VI took the SVC's advice since, in banning anabolic beef hormones, it had ignored the same committee.

In this whole affair, what is particularly disturbing is that DG VI is obviously trying to undermine the credibility and authority of the CVMP—the legally constituted regulatory review body and one which as it happens, reports to another part of the European Commission, DG III (responsible for internal market and industrial affairs). Faced with this sort of internal Commission power struggle and the resultant manipulation of the facts, manipulation of the European Parliament, and manipulation of the Council of Ministers, how can consumers or industry have faith in the regulatory system? Why, particularly, should the 26 animal health companies and 9 industry associations in FEDESA believe that other products regulated by other parts of the European Commission will be treated any more fairly than bST? This is not the first, and neither will it be the last, blow to agricultural biotechnology in the EEC. But, as a symptom of regulatory malaise, the two-year extension of the bST moratorium will have repercussions far greater than its immediate and damaging effects on those few companies developing bST.

Extract 10.3 From *The Independent*, 29 June 1991.

Will Daisy become a monster?

Dairy cows are under siege from growth hormones that could produce genetically engineered milk, says *Joanna Blythman*

If YOU were brought up with the jingle 'Drink a Pinta Milk a Day' it is likely that you went to bed at night with tales of Daisy, the contented cow, placidly chewing on lush pastures.

It is a comforting bucolic image which endures. Although many consumers now purchase milk from supermarkets in plastic half-gallon containers, the doorstep pinta, the reassuring rattle and hum of the milkman's cart in early morning streets, are symbols of a slower, more pastoral life.

In reality the modern dairy cow is quite a different animal from good old Daisy. She has been 'genetically selected to be a high performance machine, as highly tuned as a racing car… and like a car that's pushed flat out… it goes bust more easily', says Dr John Webster, Professor of Animal Husbandry at Bristol University and member of the Government's farm animal welfare council.

Other critics of the conditions of the contemporary dairy cow are even more blunt. 'It amounts to brinkmanship farming. The cow is run like a machine on the verge of a physiological breakdown, nine months of the year in milk, nine months in calf, and six months both pregnant and lactating', says Dr Alan Long, research adviser to the Vegetarian Society.

Not everyone sees it that way. 'We wouldn't accept that modern dairying is intensive or putting unreasonable strain on cows', says Geoff Smallwood of the Milk Marketing Board. He says there are signs that the average British dairy herd is getting older. But the fact remains that these days, Daisy is likely to last only into her fifth year before exhaustion and a growing catalogue of painful illnesses such as lameness, damaged udders and reproductive failure suggest that her useful life (and economic milk yield) has come to an end. By contrast, cows kept for breeding beef often live until they are 18 or 20 years old.

The welfare of dairy cows may be fairly well down the typical consumer's list of priorities. But that attitude of faith in the wholesome qualities of milk would almost certainly change if milk became a product of genetic engineering. Such a scenario and, indeed, product is waiting in the wings.

The name does not exactly trip off the tongue. Bovine somatotropin (BST) or bovine growth hormone (BGH), the more accessible name it has been given in America, is a genetically engineered hormone which has been pioneered by several large pharmaceutical companies. Injected into dairy cows at regular intervals, it can increase the cows' milk yield by at least 15 per cent. It is this product that has sparked a landmark debate on both sides of the Atlantic. Daisy the dairy cow has become the battleground for rival views of what 'progress' means in agriculture.

The debate has been focused most sharply in the US. 'The American dairy industry is in a mess. It has become so intensive that cows are suffering from complete burn-out, and that is throwing up a witch's brew of infections and viruses like bovine leukaemia and the bovine equivalent of AIDS. We've increased production so much that we've destroyed the cow's immune system', says Andrew Kimbrell, attorney at the Foundation for Economic Trends in Washington, which has been active in fighting off the introduction of BST (BGH) in the States.

While pharmaceutical companies such as Monsanto, Eli Lilly and Cyanamid pressure decision-makers to license the hormone for use, the American consumer has said no. Several supermarket chains are boycotting it.

The arguments used against BST are the same on both sides of the Atlantic. The first concern is animal welfare. BST causes an increase in mastitis — painful bovine infection — and suffering to the cow from swellings caused by frequent injections. (The American Food and Drug Administration accepts that BST has negative consequences for cows.)

Then there is the question of the desirability of increased milk yields. Quota systems were imposed in the European Community in 1984 to control over-production of milk. A milk glut would be in nobody's interests: surpluses are already running at 14 per cent in the EC.

Yet another huge imponderable is what impact BST, absorbed from the mouth or intestines, might have on human health. 'These consequences have not been researched...why allow pharmaceutical companies to impose their product on us when they haven't answered the major questions?' Mr Kimbrell asks.

In Britain, as yet, BST has not been approved by the Government's veterinary products committee, despite the apparent commitment to the product by John Gummer, the Agriculture Minister. He has said: 'We have not got to be Luddite. The idea that Britain should stand aside while allowing everyone else to produce milk in the modern way is barmy... Nobody has any doubts about damage being done to human beings, it is totally safe'. But sufficient doubts have prevailed to ensure BST has no product licence.

The European Commission has imposed a moratorium on BST until the end of this year, pending a full report on all scientific, economic and social factors. To date in Britain, BST has made few allies. In an NOP opinion poll in 1988, 83 per cent were opposed to the product. Numerous National Farmers' Union branches have stated their opposition and the Milk Marketing Board is adopting a 'wait and see' position while acknowledging that it is debatable whether farmers are receptive.

'BST would be a disaster for dairying', says Dr Webster. He believes that although BST is neither in the interest of the consumer or the agricultural industry, it might well be imposed on the dairy cow by the pharmaceutical lobby. 'If BST was licensed, farmers couldn't afford not to use it.'

But with an estimated £0.5 billion investment, in BST to date, and a tantalizing market of up to $1 billion sales per year predicted worldwide, it looks unlikely that the powerful pharmaceutical lobby will give in without a struggle. Those efforts may eventually backfire, because BST has catalysed a wider debate about the direction modern agriculture should be taking.

'Is it to mean more production in less time, regardless of its impact on the farmer, the cow or the consumer, or should we redefine progress to mean sustainability — a farming system that takes on board the environment, human and animal health and welfare?' asks Mr Kimbrell.

In other words, do you want your milk to come from Daisy or a genetically engineered and manipulated machine?

As Andrew Kimbrell has remarked about the broader implications of progress in agriculture:

> *Is it to mean more production in less time, regardless of its impact on the farmer, the cow, or the consumer, or should we redefine progress to mean sustainability—a farming system that takes on board the environment, human and animal health and welfare?*

Clearly, we now have, or will very soon have, the technology to produce new and better varieties of both plants and animals, and to get undreamed-of amounts of rare pharmaceuticals. Will financial considerations dictate the way we approach this? Will we be able to exercise a little wisdom in the take-up of this technology? We leave you to reflect on this.

Summary of Chapter 10

1 The world is fed by agricultural products. An *increase in population* requires an *increase in the amount of food produced. Conventional breeding techniques* have already supplied more and better products, but the huge prospective increase in population will require more than conventional techniques can deliver.

2 Attempts are being made to use *genetic engineering* to improve the *geographical range of plants*, their *independence of fertilizer*, and their *resistance to pests and diseases.*

3 The introduction of '*antifreeze*' genes into crop plants, and the use of *ice-minus bacteria* to coat their surfaces, may allow the plants to be grown in areas previously too cold for them.

4 The transfer of genes involved in *legume nitrogen fixation* into other species, and getting them expressed in these other plants, looks too difficult for the time being.

5 Some plants have been made resistant to insect pests by the insertion of genes coding for *B.t. toxin.*

6 One approach to *weed control* has been to insert genes coding for *herbicide resistance* into crop plants. This means that high doses of herbicide can be applied without harming the crops.

7 *Rice* is being worked on to make it *virus-resistant*, to make its *protein more nutritious*, and to make it a *source of β-carotene*, the vitamin A precursor.

8 *Cotton* is being made *herbicide- and pest-resistant*, and research is in progress to improve the *quality of the cotton fibres.*

9 Techniques for *manipulating embryos* allow clones of animals to be produced. 'Elite' embryos can be implanted into inferior, surrogate mothers, to *upgrade herd quality.*

10 It is technically difficult to get foreign genes stably inserted into animal embryonic cells, although it can be achieved.

11 In sheep, attempts are being made to improve *wool production* by *genetically engineering the cells of the sheep's rumen.*

12 *Pharming* can allow the cheap and large-scale production of valuable *pharmaceuticals* in agricultural plants and animals.

13 The application of genetic engineering techniques to agriculture, and the use of products obtained via genetic engineering (such as *bovine somatotropin, BST*), is still controversial.

Question 10.1 List some of the difficulties which had to be overcome before wheat could be genetically engineered.

Question 10.2 As much as 0.1% of the total cell protein of genetically engineered cotton plants can consist of B.t. toxin. Do you think the production of this much toxin is a good thing?

Question 10.3 What steps would you take to produce cotton plants that were resistant to cotton bollworm (assuming that this is susceptible to B.t. toxin), and get the resulting plants to market? Give your answer in note form, considering scientific, commercial and environmental aspects.

Question 10.4 In pharming, how would you ensure the tissue-specificity of drug synthesis in a host organism?

Epilogue: predicting the future

Genetic engineering is a vast topic and a rapidly changing one, with new techniques and applications seeming to appear almost daily. This obviously poses a problem in writing any book in this area. We have tried to tackle this in two ways.

On the technical side, we have restricted our discussion to just some main principles underpinning some of the major techniques. Though many variations on the themes discussed exist already and, doubtlessly, many more will be developed, these principles themselves should remain largely intact.

As for applications of genetic engineering, we have had to be very selective, concentrating our attention on two main fields—medicine and agriculture—and, indeed, on particular items within them. Inevitably, this has meant ignoring other interesting applications, both actual and potential. We could have discussed the possible use of genetically engineered microbes for cleaning up environmental pollution or for metal mining, for example. Equally, we might have dealt with the use of genetic engineering in food processing or, more speculatively, for oil recovery. However, in dealing with medicine and agriculture, we believe that we have chosen areas of considerable importance and ones where genetic engineering is set to have a major impact in the not too distant future. The principles of the techniques that you have learnt, principally from Chapters 3–8, underpin many of the developments in our chosen areas as well as potential ones, in pollution clean-up, metal recovery or whatever, that we decided to ignore.

We have also been at pains to show that applications to everyday life do not depend on technical developments alone. Social, financial, ethical and political considerations all have their part to play. Genetic engineering itself poses interesting questions for all of these domains. Reciprocally, such factors can affect considerably the progress of genetic engineering. For example, if the genetic engineering of mammals and their patenting raises awkward ethical questions, then how will the answers arrived at affect the use of genetic engineering? Will the answers and political decisions based on them encourage developments or slow them? Like the techniques and applications themselves, their relevance to such wider social and political issues is quite recent and continually developing. But, once again, what you have studied so far should provide a framework for considering and discussing such issues as they arise in the press, on radio or television, or wherever.

The potential breadth and impact of genetic engineering make any overall evaluation of its future very difficult. Undoubtedly many exciting, perhaps revolutionary, advances may occur. But what, when and how big an impact is often hard to tell—we're back to where we began in Chapter 1 with the problem of prediction. Though still difficult, it is *relatively* easy to approach prediction piecemeal; to evaluate the likely impact of each application as it is presented, something we have tried to encourage in this book. At least we hope that we have indicated the sorts of questions to be asked—technical and wider ones. As to predictions about the future of genetic engineering *as a whole*—massive impact widely felt or relatively damp squib, untrammeled benefits or moral dangers—we leave those to others.

In summing up the future of genetic engineering, I am reminded of the story about the assembling of a magazine article a few years ago. To be entitled 'What the world would like for Christmas', it was based on asking prominent figures what they would like for Christmas given a totally free choice. Reagan chose peace on Earth, Gorbachev sought the end of the arms race. A British diplomat felt quite unable to respond, but eventually ventured his choice—'A small box of crystallized fruits, please'. Though genetic engineering is certainly a powerful technology with wide applications, I should hesitate to predict it bringing peace on Earth. But of one prediction I am fairly sure: it will produce applications much less modest, and considerably more exciting, than a small box of crystallized fruits.

Further reading

Over the past 10 or so years, many books have been published on all aspects of genetic engineering, and biotechnology, at a variety of levels; doubtless many more will be. Given below is a selection of books that you may find interesting and useful to look at should you wish to learn more about these topics. In addition to books, frequent mention of advances in genetic engineering is made in *New Scientist*. A more commercially oriented, and somewhat more technical, journal that you may wish to consult is *Biotechnology*, published monthly by Macmillan. *Trends in Biotechnology*, published monthly by Elsevier, is also well-worth looking at, should you wish a somewhat more technical level than in *New Scientist*.

Books on the techniques of genetic engineering

Old, R. W. and Primrose, S. B. (1989) *Principles of Gene Manipulation: An Introduction to Genetic Engineering*, 4th edn, Blackwell, Oxford.

A well-written and more advanced book giving considerable detail on the mechanics of genetic engineering; it is good on the pros and cons of various techniques.

Singer, M. and Berg, P. (1991) *Genes & Genomes*, University Science Books, California and Blackwell, Oxford.

A large, well-constructed and well-illustrated book on all aspects of the molecular structure of genes, their functioning and manipulation; it is not purely on genetic engineering as such.

Watson, J. D., Gilman, M., Witkowski, J. and Zoller, M. (1992) *Recombinant DNA*, Scientific American Books, W. H. Freeman and Co., New York.

A well-written, well-illustrated book at a level above that of this Course.

Books on the applications of genetic engineering

Marx, J. L. (ed.) (1989) *A Revolution in Biotechnology*, Cambridge University Press.

A good multi-author book covering various aspects of the commercial/medical applications of genetic engineering and associated biotechnology. The topics include: chemicals and biochemicals from microbes; genetic engineering and medicine; genetic engineering and plant breeding. The level of treatment is appropriate for this Course.

Walgate, R. (1990) *Miracle or Menace? Biotechnology and the Third World*, The Panos Institute.

A very interesting and highly readable book on the applications of biotechnology (including genetic engineering) in the Third World.

Books dealing with social and ethical questions surrounding genetic engineering

British Medical Association (1992) *Our Genetic Future. The Science and Ethics of Genetic Engineering*, Oxford University Press.

The report of a BMA working party set up to 'consider the implications of genetic engineering'.

Kevles, D. L. and Hood, L. (eds) (1992) *The Code of Codes*, Harvard University Press.

An up-to-date multi-author book covering the technology, history and social and ethical issues concerning the Human Genome Project. The authors are major players in the areas covered.

Suzuki, D. and Knudson, P. (1990) *Genethics. The Ethics of Genetic Engineering*, Unwin Paperbacks.

A personal view of the ethical issues arising from genetic engineering. Well-written and thought-provoking.

Skills

In this section we list skills that have been explicitly taught and/or revised in this book. You should find that most of them are special instances of the general skill categories listed in the *Course Study Guide*. As usual some, such as 5 and 6, are rooted in the particular content of this book.

1 Interpret and manipulate data presented in the form of text, tables, graphs and diagrams. (*Questions 3.3, 5.5, 8.2 and 9.3; Activities 2.1 and 3.1*)

2 Summarize, in writing, the main points from a section of text that you have studied and give examples from that text that illustrate those points. (*Questions 9.4, 9.5, 10.1, 10.3 and 10.4; Activities 2.2 and 3.3*)

3 Extract information from a section, or different sections, of text that you have studied and use it to support, or refute, a given statement. This may merely take the form of categorizing the statement as true or false *or* involve presenting a longer written argument. (*Questions 2.2, 3.4, 4.1, 4.3, 4.4, 5.6, 6.3, 7.1, 7.2, 8.4 and 9.2; Activities 3.2, 3.4 and 9.2*)

4 Extract from an article information that is relevant to a particular question and consider what may 'lie behind the story'. (*Activities 8.1, 9.1 and 10.3*)

5 Design, in principle and outline, an experimental procedure(s) to achieve a particular given end *or* predict or comment on the outcome of a given procedure. (*Questions 4.5, 4.6, 6.1, 6.2, 8.1, 8.2, 8.3, 8.5 and 10.3; Activities 4.1, 6.1, 6.2, 7.1 and 10.2*)

6 Given a proposed project, real or hypothetical, suggest the criteria—technical and other—that would help determine the success or failure of the project. (*Questions 8.6 and 10.1; Activities 7.1 and 8.1*)

7 Consider social, political and ethical aspects of a scientific issue. (*Activities 8.1, 9.2, 9.3 and 10.3*)

8 Formulate a personal opinion on a scientific issue. (*Activities 9.2 and 10.1*)

Answers to questions

Question 2.1

The completed passage is as follows (with the words added printed in italics).

A species of rodent has a *haploid* number of 14, this being the number of *chromosomes* in its gamete cells. Its *diploid* number is therefore 28, this being the number of *chromosomes* in its *somatic* (i.e. body) cells. Varieties of the rodent are known, varying in ear length and/or eye colour. There is a *gene* for each of these two characteristics and these two *genes* (for ear length and eye colour) exist on different *chromosomes* and so assort *independently* during *meiosis* (i.e. the production of gametes). Two different *alleles* of the *gene* for ear length are known; one responsible for long ears, the other for short ones. Likewise there are two different known *alleles* of the *gene* for eye colour; one responsible for red eyes, the other for white eyes.

Question 2.2

(a) False. The genetically identical eggs would still have to undergo the highly random process of fertilization. Even if the extremely unlikely event of the production of two genetically identical sperm occurred, it would still be very unlikely that these two sperm, among very many, would fertilize the two identical eggs among many.

(b) False. The four taxonomic kingdoms comprise: animals, plants, fungi and prokaryotes. Genetic engineering can cross between them—for example, bacterial genes into plants (end of Chapter 2).

(c) True (end of Chapter 2).

Question 2.3

Among the very many to chose from are: pedigree racehorses, cats, dogs; farm animals; various crop plants such as wheat, rice, potato; garden flowers.

Question 3.1

All are eukaryote except *E. coli*, which is prokaryote.

Question 3.2

Scotch whisky, bread and penicillin are all products obtained using microbes at some stage or other in the process. In *some* senses this is also true of oil which is in part the product of millions of years of microbial action on dead organisms (including microbes)—it is debatable whether this can strictly be called exploitation of microbes in the same sense as whisky, etc. (However, microbial products have been used also, sometimes just experimentally, in *oil recovery*, helping force crude oil out from under the ground.)

Question 3.3

Neither; they will yield the same amount. There is no need to calculate how many cells will be present—the cells in A divide at half the rate of those in B but have twice the time to divide; the net result is the same number of rounds of division (three) and hence the same final number of cells.

Question 3.4

(a) True (assuming the bacteria in question are useful in other ways; e.g. give a useful product).

(b) True *or* false. The need for a simple growth medium is attractive as this simplicity probably means 'easy and cheap to prepare'. However, the ability of virtually any bacteria to grow in the medium implies that careful precautions will have to be taken to exclude all but the desired bacteria—i.e. maintain sterility (Section 3.1). On an industrial scale this can be difficult and expensive; whether this is acceptable will depend on how expensive and how valuable the product is. The commercially ideal situation is a cheap growth medium which is peculiarly suited to just the bacteria of interest.

(c) True (Section 3.4).

(d) True so far, though this *may* change with time (Chapter 1 and Section 3.4).

Question 4.1

(a) False. Other genes may encode other information; for example, some genes encode information for making rRNA which is *not* translated to give a polypeptide chain.

(b) True. Eukaryote chromosomes contain DNA and proteins, of which the histones are an important group.

(c) False. Chromosomes are indeed large structures but it seems that each chromosome contains just a single very long molecule of DNA.

(d) True. Each chromosome contains just one molecule of DNA but this long molecule contains many distinct genes within it.

(e) True. There are different types of deoxyribonucleotide depending on which one of four (deoxyribonucleotide) bases each contains. Thus the sequence of bases 'spells out' the information in a gene.

(f) False. There are indeed many genes within a single large molecule of DNA but the delineation between genes is in the form of 'punctuation' rather than by breaks in the molecule. That is, sequences of bases act as 'punctuation' indicating the beginnings or ends of genes.

Question 4.2

The sequence in the other, complementary strand is:

TTGCCCAATGGACATGGTTTA

Question 4.3

(a) False. It is deoxyribose.

(b) True.

(c) False. Adenine pairs with thymine, and cytosine with guanine.

(d) False. It is somewhat difficult to present 'what would happen *if* ...' type arguments in evolution as we cannot 'change history'. Nevertheless, separation of the two strands of the DNA double helix is an essential step in DNA replication and this is easy because of the weakness of the hydrogen bonds between base pairs. So if these were covalent bonds instead, it would be more difficult to separate the strands. Thus it is at least reasonable to speculate that such a molecule would be *less*, not more, efficient as a heritable carrier of information.

Question 4.4

(a) (i) only—it is composed of DNA. Prokaryotes have no histones nor any nucleus.

(b) (i) only—see (a).

(c) (i), (ii) and (iii).

Question 4.5

Not all steps were given but of those that were the correct sequence is (a), (d), (b), (c)—see Figure 4.2.

Question 4.6

(i) (d)—to effect the splicing.

(ii) (a)—Section 4.3.

Question 5.1

The sequence of the 10 amino acids made by translating this stretch of mRNA is:

Met Leu Tyr Phe Trp Gly Cys Val Phe Leu...

Question 5.2

The base sequence of an mRNA must be complementary to that in the *coding* strand of the DNA of the gene from which the mRNA is transcribed. Read from left to right, the sequence complementary to the portion of mRNA shown is:

...TAC AAC ATA AAG ACC CCT ACA CAA AAA AAC...

(*Note*: T not U, as this would be in DNA.)

In turn, the other, non-coding, strand of the DNA must be complementary to the coding strand. Therefore the base sequence (in both strands) of the relevant portion of the gene must be:

...ATG TTG TAT TTC TGG GGA TGT GTT TTT TTG...
...TAC AAC ATA AAG ACC CCT ACA CAA AAA AAC...

Question 5.3

If the 15th base in the mRNA were A, this would convert the fifth codon to UGA. This is a stop codon and hence the reading (translating) of the message would stop here. Therefore the resulting (poly)peptide would be truncated and be Met Leu Tyr Phe *in total*.

Question 5.4

Being a prokaryote, *E. coli* does not have the necessary machinery for post-transcriptional modification of (primary) newly produced mRNA transcripts of genes with introns. Even if such genes are transcribed in *E. coli*, without such post-transcriptional modification the mRNAs cannot be translated to give the normal protein products of the genes. As *prokaryote* genes, those in (a) and (c) will not have introns, and neither does the eukaryote one in (b). Therefore the genes described in (a), (b) and (c) might all be expressed to give their normal protein product. (*Note*: the question only considered the problem of split genes; it says nothing of other reasons—for example, incompatible promoters—that might actually prevent translation of, say, the mRNAs from (b) and (c) occurring.) Those in (d) and (e), having introns, could not—the number of introns is irrelevant, as all those portions of the mRNA corresponding to introns must be excised before the normal protein product of a gene can be produced.

Question 5.5

(a) A, RNA polymerase; B, mRNA; C, ribosome; D, growing polypeptide chain; E, polypeptide.

(b) All of them.

(c) Both items. In eukaryotes DNA is in the nucleus while the ribosomes are in the cytoplasm. A *nuclear envelope* separates the nucleus from cytoplasm and must be *present* always as, by definition, the presence of a distinct nucleus is a key feature that distinguishes eukaryotes from prokaryotes. It *operates* in the pathway in the sense that mRNA must traverse it to reach the cytoplasm. *The machinery of post-transcriptional modification* will always be *present*. It operates where the genes being transcribed are split ones, cutting and splicing the primary mRNA transcripts of such genes, thereby removing intron-corresponding sequences (Figure 5.4). But not all eukaryote genes are split ones, and in such cases the primary mRNA does not need this post-transcriptional modification.

Question 5.6

(a) False. If we merely wish to transfer the genes and are not concerned about their expression there is no need to remove any introns from the DNA (also, in some eukaryote genes there may not be any introns, anyhow). Indeed, should we wish to transfer a *complete* gene, including introns, cDNA cloning could *not* be used as a cDNA is effectively a 'gene minus introns' (see later, in Section 7.1).

(b) True or false; it depends on the gene in question, whether it is split or not. Most, but not all, eukaryote genes are split. If it is split, then cDNA cloning would be appropriate. If unsplit, then either method, cDNA cloning or cloning of gene-size fragments from restriction enzymes, might be used. (However, we may only find out by trial and error whether a particular gene is split or not. As the cDNA works in either case, it is the method to choose, if otherwise appropriate, e.g. a good source of mRNA is available.) In either case, we must also ensure that other things, such as a compatible promoter, are present.

Question 6.1

(a) (i), (ii) and (iv). (i) is needed to cleave foreign DNA and to open plasmid circles (Figure 4.2d and b); (ii) is needed to get uptake of recombinant plasmids into bacterial cells (Figure 4.2f); (iv) is needed to splice foreign DNA fragments into opened plasmid circles (Figure 4.2e). (iii) is *not* needed.

(b) (i), (ii), (iii) and (iv). (i) is needed to obtain opened plasmid circles (Figure 5.6c) but is *not* needed to cleave foreign DNA; (ii) is needed to get uptake of recombinant plasmids into bacterial cells (Figure 5.6d); (iii) is needed to produce cDNAs from mRNAs (Figure 5.6b); (iv) is needed to splice foreign DNA fragments into opened plasmid circles (Figure 5.6c).

Question 6.2

Either technique can be used. An enzyme can be detected by virtue of its specific catalytic activity *or* by its specific binding to an antibody against it. In contrast, for a non-enzyme protein only the antibody method would be appropriate.

Question 6.3

(a) False. With very few exceptions prokaryote genes have no introns, so a cDNA is a faithful copy of a (mature) mRNA which is itself a faithful copy of a structural gene. So the cDNA and the gene it represents should be identical in length.

(b) False, in general. A cDNA library represents the total population of types of mRNA being made by a particular type of cell. In general, not all the mRNAs being made in a cell are in equal amounts; mRNAs for abundant proteins in that cell type are themselves abundant (Sections 5.3.2 and 6.1). Thus, when the mix of cytoplasmic mRNAs is extracted from a cell some mRNAs will be more abundant than others. When copied to give a mix of cDNAs, this will therefore contain some cDNAs in greater amounts than others. When then cloned (Figure 5.6c–f), clones containing the relatively abundant cDNAs will turn up more often. Hence the number of clones containing each *different* type of cDNA will not be equally represented.

Question 7.1

(a) False. It can be used as a vector for either. The point about λ is that it can tolerate large inserts of DNA; the source of the DNA—restriction enzyme fragment or cDNA (or synthetic gene)—is irrelevant.

(b) True, but not all proteins. However, some tricks exist for enabling desired foreign proteins to be secreted.

(c) It rather depends on what is meant by 'correctly'. Yeast can attach sugars to their proper places on eukaryote polypeptides but probably does not generally attach the same sugars to foreign polypeptides as would be attached in their native cells.

(d) False. They are alright as hosts provided that care is taken to purify the foreign protein product free of cell wall toxin. However, such purification may be difficult and therefore expensive; it might then make better sense to use an alternative host.

(e) False. cDNA is prepared by copying *mature* mRNA which has no base sequences corresponding to introns. Thus a cDNA will lack such intron sequences. On the other hand, the restriction enzyme technique would be applicable, as it operates on the actual DNA of the genome where both any introns and exons in a gene will be present. (Of course, if the restriction fragment thus obtained is very large, this will present another problem—getting a suitable vector.)

Question 7.2

(a) Neither; generally. *E. coli* does not add sugars to polypeptides and, assuming the foreign gene is split (as most, but not all, eukaryote genes are), the gene with its introns still in place cannot be properly expressed.

(b) (i) is true *but* any sugars attached to foreign proteins may not be the same as those (if any) normally attached in its native cells. (ii) is not true. You may have concluded reasonably that, as a eukaryote which itself contains split genes, yeast should be able to carry out post-transcriptional modification of mRNA from split genes of other eukaryotes. Although yeast does remove introns from its own primary mRNA transcripts, it does not (in general) correctly cut and splice mammalian transcripts. So it may be that when using yeast as a host for mammalian proteins, it is generally advisable to use the cDNA route.

(c) (i) is true *but*, as with yeast, there is no absolute guarantee that mammalian tissue culture cells will put the correct sugars on the polypeptide. (ii) Such evidence as exists suggests that, as we might expect, mammalian tissue culture cells can process correctly primary mRNA transcripts of foreign mammalian genes.

Question 8.1

Obviously the foreign gene must be spliced into the T_i plasmid somehow and presumably this could be achieved by the usual splicing methods outlined in Chapter 4 (see Figure 4.2, for example). The *recombinant* T_i plasmid could then be used to carry this foreign gene into the plant cell as desired.

However, we also want this gene integrated stably into the plant chromosomal DNA. Remember that only a part of the T_i plasmid DNA ends up so integrated—the T-region. So the trick would be to splice the foreign gene into the T_i plasmid *within* the T-region.

Question 8.2

(a) Remember that the gene for rat hormone was transferred complete with its five introns, not a cDNA. As growth hormone seems to have been produced successfully, the post-transcriptional modification machinery in the host, the mouse, must be very similar to that in the source of the gene, the rat. The same argument holds for the human growth hormone gene which, as you probably assumed, also contains introns.

(b) In a number of instances the rat or human growth hormone seems to have produced 'giant' mice, suggesting that a growth hormone is not totally species-specific. Rat or human growth hormone is biologically active in mice.

Question 8.3

In the liver and kidney. In plants, regulatory DNA sequences are tissue-specific; that is, it is the adjacent regulatory sequences that govern in which tissues the synthesis occurs, not the structural gene itself. The regulatory sequences used come from adjacent to the metallothionein gene and ensure metallothionein synthesis in the kidney and liver. So, they should function similarly when placed alongside rat or human growth hormone gene.

Some experiments using regulatory sequences adjacent to the metallothionein gene and other regulatory sequences have indicated that tissue specificity can be obeyed.

Question 8.4

(a) False; the converse is generally true.

(b) True.

(c) False; they are made in *plant* cells, not in *A. tumefaciens*, but only in plant cells that have been changed by the action of *A. tumefaciens*, i.e. by integration of T-DNA.

(d) False; integration is generally at random sites.

Question 8.5

(a) The most likely explanation is that the donkey regulatory DNA sequences spliced to the gene for protein U were brain-specific.

(b) This implies that donkein is normally made in the donkey brain.

Question 8.6

The *E. coli*-based one. Techniques for transferring foreign genes or their cDNA equivalents into *E. coli* and getting their expression there are well advanced (Chapters 4–6). The wheat project is much more fraught with technical problems. Obtaining the bacterial gene should not be a major stumbling block but successfully transferring it into the cells of wheat will be much more difficult than with *E. coli* as a host. There is also the very considerable problem that wheat cannot be easily regenerated from single cells or explants (Section 8.4). An alternative route, of the biolistics variety (Section 8.4.1), has proven usable in principle with some plants, including a member of the Graminae (rice), but, at the time of writing, wheat has only just succumbed to this approach. So the wheat project would likely be very difficult with present techniques.

Naturally the absence of a suitable technique, or using a difficult one, presents a stumbling block to an application. However, on the other hand, if the motivation for the application is strong enough, this in itself can provide a powerful spur for development of new techniques. The desire to cure or prevent dreaded diseases such as cancer or AIDS, or to improve an important crop plant, are obvious examples where motivation has led to considerable scientific research, and hence innovative techniques, in genetic engineering included (e.g. the OncoMouse; and see Chapters 9 and 10). Nevertheless, other things being equal, projects where totally new techniques or fundamental knowledge must be acquired first will inevitably take longer to reach completion than those chiefly employing existing know-how.

Question 9.1

The ability to shuffle DNA, producing rearrangements coding for antibodies (Figure 9.2) is a feature unique to (ancestor) spleen cells. Bacteria are highly unlikely to be able to do this. We could, however, envisage transferring *already rearranged DNA* (i.e. coding for a specific antibody polypeptide chain), isolated from one clone of spleen cells, into bacteria. There would still be the problem that this rearranged DNA contains introns (Figure 9.2), which bacteria cannot generally cope with (Box 5.2).

Question 9.2

(a) True; it masks the white characteristic in all the offspring.

(b) False; they have all inherited one dominant yellow allele and one recessive white allele from their true-breeding parents, so they are heterozygous.

(c) False; any offspring containing two recessive alleles would be phenotypically recessive, i.e. white.

(d) False; All the F_1 contain one dominant yellow allele and one recessive white allele. So when they breed, they can pass on either the white or the yellow allele. This means that *their* offspring (i.e. the F_2 generation) can inherit two dominant yellow alleles, one dominant and one recessive allele, or two recessive white alleles. Because yellow is dominant to white, the proportions of each colour in the F_2 will be three yellow to one white.

(e) True; see above.

Question 9.3

(a) Affected individuals must have two copies of the defective allele, because it is recessive. In generation II, all the offspring had one defective allele and one normal allele. They were therefore carriers, but did not suffer from the disease themselves. The individuals in generation II *all* inherited the defective allele. This was unlucky, as they each had a one in two chance of *not* inheriting it.

(b) In generation VIII, only the individuals carrying two copies of the defective allele would be affected by the disease. All the parents carried one copy of the defective allele, so according to the laws of probability they all had a 50% chance of passing on that allele to any of their offspring. Thus, in theory, we would expect 25% of the children to inherit a defective copy from *both* parents, and therefore suffer the disease, 50% of the children to inherit one defective copy from one parent, plus a normal allele from the other, and therefore be carriers, and 25% of them to inherit two normal alleles (see Figure 9.6). This last group would be disease-free, as would have been their children. In fact there were 33%, not the theoretical 25%, affected individuals—this discrepancy is not surprising, considering the small numbers involved.

Question 9.4

Advantages

Subunit vaccines can be produced easily by genetically engineered bacteria, and they are safer than conventional vaccines because there are no complete pathogens involved in the mass production process.

Disadvantages

Subunit vaccines tend not to elicit such a strong immune reaction as do whole pathogens. They are also rather expensive to produce at present.

Question 9.5

Genetic engineering has allowed the production of enough pure interferon for large-scale clinical trials to be carried out to determine its effectiveness as an anti-viral and anti-cancer drug, and, if found to be useful, for therapy itself.

Question 10.1

Having identified genes which would be useful in wheat (e.g. herbicide resistance), the main difficulties in genetically engineering the wheat were as follows:

1 To get the foreign DNA into wheat cells.

2 To get the genes expressed in their new host.

3 To regenerate wheat plants from tissue cultures.

These problems are obviously common to all plant genetic engineering work.

Question 10.2

Yes and no. The toxin may need to be present in high quantities to be effective (as was the case for the cowpea protease inhibitor, see Section 10.2.3), but the energy demands of the extra protein synthesis may be enough to reduce the overall cotton yield (as might well be the case for nitrogen fixation, see Section 10.2.2).

Question 10.3

Your answer should have made the following points:

1 The gene for B.t. toxin (which has already been isolated) should be put into cotton cells, by using the T_i plasmid as a vector or by biolistics.

2 The cells should be grown into whole plants via tissue culture. This may be a lengthy process.

3 Several generations of plants must be bred and tested in field trials for resistance to bollworm. Objections of environmentalists to these field trials may be encountered.

4 You should demonstrate that the yield of fibres per plant is at least as good as normal, otherwise your plants will have little commercial advantage.

5 The plants, and the cotton derived from them, must be shown to be safe, as environmentalists may object to the use of the cotton.

Question 10.4

Tissue-specific expression can be assured by putting the pharmaceutical gene adjacent to the regulatory sequences of a gene only ever expressed in the tissue to be used.

Answers to activities

Activity 2.1

(a) The wheat yield in 1950 was about 2.7 tonnes per hectare (i.e. the value at the point on the vertical axis corresponding to the graph line at 1950). In 1970 it was around 4.2 tonnes per hectare. This is an increase of 1.5 tonnes per hectare; that is, $1.5/2.7 \times 100\% \approx 56\%$. Therefore there was about a 56% increase in yield over those years.

(b) No. To do so you also would need to know the *total* area of land under cultivation for wheat in 1975 and this is not given by the graph.

(c) The graph undoubtedly demonstrates that wheat yield increased considerably between 1940 and 1980, around a doubling in tonnes per hectare. Thus these data are *consistent* with the claim (hypothesis) that breeding of new plant varieties has led to improved yields. However, we cannot take the data as strictly demonstrating the validity of the claim because many factors other than new wheat varieties may have also contributed to increased yields.

Activity 2.2

Summary of main points of Chapter 2:

1 Human beings have long sought to improve crops and farm animals by means of *selective breeding*—i.e. choosing desirable animals/plants from which to breed. (Crops; dog breeders.)

2 Plant and animal breeders have been quite successful, notably since the advent of genetics. But the randomness of inheritance and the species barrier have both often frustrated their efforts. Randomness of inheritance is due to random mutation, recombination and random fertilization.

3 The randomness of inheritance means that one *cannot guarantee* that the offspring of two chosen parents will have the *desired mix* of parental characteristics. Breeding is thus to some extent a *gene lottery*. (Isadora Duncan/Bernard Shaw)

4 *Genetic engineering* is a set of techniques that allows us to *transfer specific genes* from one organism to another. It *by-passes the gene lottery* of sexual reproduction; just the desired gene(s) is transferred.

5 Genetic engineering can be used between different species, even essentially unrelated ones. It thus *breeches the species barrier*. (Bacterial herbicide gene into tomato; gene for human blood clotting factor into sheep; blue rose.)

6 Overall, genetic engineering allows a degree of *precision* in which genes are transferred and from which species into which other, that is unachievable by classical breeding techniques.

Obviously there is no one 'correct' summary, but that given above does summarize quite concisely the main points of Chapter 2. There are a few other things to note about it:

(i) The order of a summary need not be strictly that of the material it summarizes—this chapter starts with a statement of what genetic engineering is, followed by a question of why anyone should be interested, leading to a discussion of traditional breeding techniques, followed by what genetic engineering has to offer uniquely. In the above summary the order is very slightly different—we start with the interest in traditional breeding techniques *then* build up to what genetic engineering has to offer uniquely. You will note that the last item of the summary (6) in effect summarizes the summary itself, thus bringing out the central power of genetic engineering (to some extent what the *chapter* mentioned at the beginning). You could retain the original chapter order in the summary, just as well, if you wished.

(ii) We have emphasized key words and phrases by italicizing them. You may find it useful to underline or highlight such items, should you wish to return to your summary and quickly pick out certain points.

(iii) The summary was intended to be of the main points *not* of the illustrative material. However, it might be useful to note down briefly some examples of these main points and this we have done in parentheses. This can be useful in two ways: as an aid to memory (for example, 'buzz words' like Isadora Duncan might help recall the main point that sexual reproduction involves a 'gene lottery'); and should you need later to present an argument involving any of the main points, it may be useful to have a brief note of examples you could use to illustrate them. (You could, if you wish, also include references to page or section numbers where the points or illustrations are made most forcefully—build up your own 'index'.)

Activity 3.1

(a) As the cycle of growth and division takes 40 minutes, each *E. coli* becomes two *every* 40 minutes. So after just 40 minutes, the original single *E. coli* becomes two; after 80 minutes these two become four; and so on. Growth is said to be exponential and the progression can be expressed by powers of two—i.e. at time zero there are 2^0 (i.e. 1) cells, after 40 minutes there are 2^1 (i.e. 2) cells, after 80 minutes there are 2^2 (i.e. 4) cells, and so on. Six hours is 9×40 minutes; therefore there will be 9 rounds of cell division overall and hence 2^9 (i.e. 512) cells. After 8 hours there will be a further 3 divisions and hence $2^3 \times 2^9$ cells, that is 2^{12} (i.e. 4096) cells. These calculations give you some idea of the rapidity of growth of microbes; after just 8 hours there is over 4000 times the starting number of cells.

(b) Six hours with a 40 minute cycle yielded 512 cells. To get 512 (i.e. 2^9) cells from one required 9 divisions. Thus with a doubling time of 30 minutes, this will take 9×30 minutes, i.e. $4\frac{1}{2}$ hours. Comparing this with the answer to (a), you can see what a considerable saving in time is achieved by a faster doubling rate.

(c) 20 days = $20 \times 24 \times 60$ minutes. As the doubling time is 40 minutes, *in principle* there will be $20 \times 24 \times 60/40$ doublings = 720 doublings. So in 20 days one *E. coli* (i.e. one cell) would, in theory, give rise to 2^{720} cells. This is approximately 10^{217} cells.

Activity 3.2

1 Breeding new varieties of animals, plants or microbes all depends on having *genetic variation.*

2 All genetic variation depends *ultimately* on availability of mutants.

3 *Where* in the DNA mutations occur is *random*, in animals, plants *and* microbes. *Specific* mutations cannot be caused to occur (induced); desired mutants, if present, can only be selected afterwards from the ones *randomly* produced.

4 In animals and plants, sexual reproduction also leads to more variety by throwing up *new combinations* of existing alleles (mutants). But this variety cannot be specifically controlled (gene lottery).

5 In asexual reproduction, as in hypothetical microbe, no problem of gene lottery of sex. But *still* randomness of where mutations occur. Therefore which mutants (alleles) become available *still* random (point 3 above).

6 Conclusions. Asexually reproducing microbes — get more mutants quickly. Therefore more to chose from. But still *randomness* of what mutations occur; *cannot* specifically control which mutants occur any more than in plants and animals. But no gene lottery of sexual reproduction; therefore breeding identical offspring from any useful mutants that do *happen* to arise in asexual microbes is, in some sense, more 'controllable' by the experimenter.

Activity 3.3

A summary of the main points of Chapter 3 is given below.

1 *Microbes* (also called *micro-organisms* or '*bugs*') are invisible to the naked eye. They include both *prokaryotes* and *eukaryotes*. Some are *unicellular*, others multicellular. Among microbes are *bacteria*, blue–green bacteria, fungi (moulds and *yeasts*) and protista. Viruses are also microbes; they are not cellular and are too small to be seen under the light microscope. In this book *generally 'microbe' refers to a unicellular organism*.

2 Microbes have long been *exploited* by humans, well *before even their existence was realized*. (Beer, bread, cheese; Table 3.1.)

3 Processes such as brewing or baking are nowadays sometimes referred to as *traditional biotechnology. Biotechnology* as a whole is the *provision of goods or services by means of living cells* and *modern biotechnology* encompasses genetic engineering.

4 In this century the use of microbes has been extended by discovering *new microbes* and developing *optimum growth conditions*. (Antibiotics, food additives — e.g. MSG, bacterial insecticide.)

5 Microbes can reproduce *asexually* and at *rapid rates*. Thus they can be grown in *bulk*. (*Saccharomyces cerevisae* — baker's yeast; *Corynebacterium glutamicum* for glutamic acid and lysine — Figure 3.3.)

6 Asexually reproducing microbes can give rise to *new strains* (i.e. varieties) by *mutation*. What mutations turn up is *random*. But the *vast number of organisms* produced makes the likelihood of *finding a desirable mutation* greater than in animals and plants where only relatively small numbers of organisms can be produced. The number of mutations occurring (still random) can be increased by using *mutagens*. (*Penicillium chrysogenum* producing 55 times amount of penicillin.)

7 A sort of *sexual reproduction* does exist in some microbes and this too can be used to produce new varieties.

8 But finding new strains via *mutant selection* or *sexual reproduction* both involve random events — one cannot control what variety (new mutant alleles or new combinations) turns up.

9 *Genetic engineering of microbes* like that of animals and plants allows *specific gene transfer* and *across the species barrier*. (Human products in bacteria — insulin, interferons; plant products in bacteria — thaumatin.)

10 Genetic engineering in microbes allows us to further exploit a long exploited feature — the ability to grow microbes in *bulk*, often cheaply. We can use this to produce more readily other *('foreign') products*, that are otherwise hard or more expensive to obtain.

Obviously your summary will not be identical to the above but it should agree with it broadly as to which are the main points. You may also have chosen different examples. We have also included a few 'index'-type references (e.g. Figure 3.3); something that, in the answer to Activity 2.2, we suggested might be useful.

Activity 3.4

(a) A logical case would be along the following lines:

1 Humans have exploited animals, plants and microbes for many years (crops, farm animals, beer, bread).

2 We have tried to improve animals, plants and, more recently, microbes by breeding new varieties (wheat; pedigree dogs, racehorses; strains of yeast and antibiotic-producing microbes).

3 Traditional breeding techniques include searching for suitable genetic variants (arising via *random* mutation) and, where sexual reproduction is available, creating new combinations of alleles.

4 Both mutation and sexual reproduction involve *random* events. This has limited the degree of control that we can have over the varieties produced.

5 The species barrier has also restricted the possible variety obtainable.

6 Genetic engineering allows transfer of just one or more *specific* genes from one organism into another—it thus by-passes the gene lottery of sexual reproduction and it also can breech the species barrier.

7 Thus genetic engineering permits greater variety (i.e. genes from different species to be incorporated into the same organism) and allows us to exercise more control, more specificity over what is incorporated—it is these two aspects that represent the *revolution* wrought by genetic engineering.

8 This (potentially) enables us to produce rare products in bulk in novel microbial 'factories' (interferons, human insulin) and to introduce specifically into animals and plants useful genes from other species (bacterial insecticide gene into plants).

(b) A logical case would be along the following lines:

1 Humans have exploited animals, plants and microbes for many years (crops, farm animals, beer, bread).

2 We have tried to improve animals, plants and, more recently, microbes by breeding new varieties (wheat and maize; pedigree dogs, racehorses; strains of yeast and antibiotic-producing microbes).

3 The efforts of plant, animal and, more recently, microbe breeders have been somewhat frustrated by nature. Namely: mutation is *random* and so (sometimes impossibly) large numbers of organisms may need to be searched to find suitable genetic variants; the variety introduced by new combinations of alleles arising via sexual reproduction is subject to randomness (the gene lottery); cross-breeding to obtain new combinations of alleles is generally restricted to within a species (due to the species barrier).

4 *Genetic engineering* can enable breeders to overcome these hurdles imposed by nature, allowing greater *specificity* over what genes are transferred and also permitting transfer of (useful) genes from one species to virtually any other.

5 Thus genetic engineering is a very specific and powerful tool; just the latest tool for the time-honoured tradition of selective breeding (from Mexican maize to more productive strains of *Penicillium chrysogenum*).

Once again, the exact form of your answers to (a) and (b) will vary. But you should have covered much the same ground as we did.

It may have seemed odd to you that we should ask you to support what at first sight seem like opposite points of view. In (a) we contended that genetic engineering *is* revolutionary. We support this in our answer by showing that *for the first time* it affords breeders a great degree of control over what genes an organism will 'inherit' and where it 'inherits' them from (i.e. which species). We argue that this offers a true revolution in the range and types of new uses to which organisms can be put. In (b) we contended that genetic engineering was *not* revolutionary, but just a logical extension of old practices. Interestingly, in our answer we covered much the same ground; we deliberately made points 1 and 2 identical to show this. But we have shifted *emphasis*—while *in effect* conceding implicitly that the *techniques* of genetic engineering are 'revolutionary' we maintain that they represent a continuation of thousands of years of practice, the selective breeding of organisms to suit human wishes. When viewed from the perspective of the *intention* behind using genetic engineering, it indeed appears as a logical extension of what humans have been trying to do for so long. Admittedly it does *somewhat* duck the case where very rare substances (e.g. interferons), produced by genetically engineered microbes, would otherwise be virtually unobtainable. This could be argued to be the weak point in this case (b). But even here one *might* perhaps claim that the general *intention* to obtain rare substances is not new as such, merely that the means has been unavailable hitherto.

Thus in summary, genetic engineering is a *revolutionary* new tool (set of techniques) with which to *extend a long-held desire* to selectively breed new varieties of animals, plants and microbes. The revolutionary aspects lie in the degree of specific control over what is inherited and the ability to breech the species barrier, allowing much greater variety to be created and more substances to be obtained; the mere extension lies in the general aims of plant, animal and microbe breeders.

It is important to see from this activity how often virtually the same data can be taken to support what at first sight seem like opposite cases. Had you been asked 'Give evidence for and against the contention that genetic engineering represents a revolution in the breeding of plants, animals and microbes' you would have had to consider *both* the *technical* revolution and the essentially long-held *aims* (i.e. breeding better varieties). You will also note how in Chapters 2 and 3 we have led gradually from the long-held aims and traditional techniques to the newer tool of genetic engineering. This should have made clear that the essential *motivations* behind using genetic engineering are long-standing. It is the power to carry out the aims much more effectively that engenders the excitement.

Activity 4.1

To cleave DNA a restriction enzyme must have its target sequence within that DNA. To cleave out a piece of DNA from within a longer molecule, as is required here, a restriction enzyme must therefore have at least two such sequences present within the molecule. An examination of the sequence data given will reveal that only two of the four enzymes named, *Alu* I and *Hae* III, have their target sequence present and both in fact have it more than once. In Figure 4.8, the target sequence for *Alu* I is identified by the pink boxes; that for *Hae* III by the grey boxes.

...TTCAGCTAGGCCCTATGCTAGCTACC... AGTTCCTAATAAGAGGCCTAAGCTTATCCTA...

...AAGTCGATCCGGGATACGATCGATGG... TCAAGGATTATTCTCCGGATTCGAATAGGAT...

Figure 4.8 For Activity 4.1 answer.

If *Alu* I were used it would cut either side of the desired gene, as required to clip it out, *but would also cut within that gene* (at the second of its target sequences reading from the left). It would therefore cut the desired gene into two fragments which would presumably be spliced into different plasmids and hence transferred independently to different *E. coli* (Figure 4.2). Unlike loaves, half a gene is *no* better than none—we need to transfer intact the gene that we want.

Using *Hae* III would create just two cuts, one either side of the required gene and both *outside* it. *Hae* III would therefore cleave the DNA leaving the desired gene intact within a gene-size piece of DNA, just as needed. (Note that in practice there would be no need to cut out the required gene quite so precisely, longer sequences of bases could be present between the cleavage sites and the ends of the gene. The important thing is that the cleavage sites lie outside the gene. I just did not want to give an even longer base sequence to examine.)

Of course, as already mentioned in the text, in practice we should rarely know the base sequence of the gene we seek in advance. So, using such an approach, 'choice' of an appropriate restriction enzyme would be more a matter of trial and error.

Activity 6.1

In principle, there are two ways in which you could have chosen to transfer interferon-producing capacity to *E. coli*. One is to clone the complete human genome using a restriction enzyme and a protocol along the lines of that in Figure 4.2. This has two disadvantages. Firstly, a very large number of different clones will be generated and therefore must be screened. Secondly, there is the problem of expression. We want the protein product (α-interferon). Even if the gene-size fragment containing the α-interferon gene has an adjacent promoter, will this human promoter be recognized by *E. coli* RNA polymerase? And there is also the problem of split genes. What if the gene for human α-interferon happens to be a split one? (As it so happens, it is not; but you were told to assume that you knew nothing of the structure of the gene in advance.) For these reasons it is therefore better to adopt another cloning strategy altogether—that of cDNA cloning.

To do this you must first obtain the mRNA corresponding to the α-interferon gene. This can be done by extracting the mixture of *cytoplasmic* mRNAs (already modified post-transcriptionally; Figure 5.4) from cells known to produce α-interferon; human leukocytes are the obvious choice. This mixture of mRNAs can be used to generate a mixture of cDNAs (Figure 5.5) among which should be the desired cDNA, that coding for α-interferon. These cDNAs can then be cloned in *E. coli* by a protocol like that in Figure 5.6c–g, making sure to use a plasmid that is an appropriate expression vector.

This is necessary as you wish not just to transfer the gene (cDNA) for human α-interferon into *E. coli* but also to get it expressed there to generate large amounts of the protein. To do this, having the gene under the influence of a *compatible* promoter is vital. This can be arranged by using a plasmid that is an expression vector for *E. coli*; in fact, a plasmid that actually has an *E. coli* promoter adjacent to the spliced-in cDNA for α-interferon.

Screening the clones to see which produce human α-interferon can be done in two quite different ways. Remember interferons are *not* enzymes, so an assay for some enzymic activity is ruled out as the basis of a screening technique (Section 6.3). Therefore, one way is to produce an *antibody* specific for human α-interferon and see which of the clones (colonies) contain protein that can bind it. Another way is to exploit the known biological activity of α-interferon, the inhibitory effect on viral growth. To do this, samples from each colony could be tested in some assay of viral growth. That is, in essence, human cells infected with a virus could be treated with a sample taken from a colony. It is then seen whether the sample inhibits the growth of the virus as compared to an untreated control (i.e. infected cells without a sample of the colony). Where inhibition of viral growth occurs, the colony from which the sample is taken must be producing interferon.

Given the information available in Chapters 4–6, the above answer is satisfactory. However, you might like to note that in fact α-interferon is synthesized in its native cells as a *precursor* polypeptide with an additional 23 amino acids at its N-terminal end. These are probably cleaved off during secretion of the *mature* α-interferon from leukocytes. The cDNA route outlined on the previous page would lead to the synthesis, in *E. coli*, of the precursor retaining the extra amino acids. Therefore, in practice, to get mature α-interferon ways round the problem of the additional amino acids have been devised. We need not consider what these are.

Activity 6.2

You were told that a hybrid polypeptide was produced. Thus the cloning of the synthetic gene *per se* and its expression seem to have worked. So the problem must lie elsewhere. To see why llamastatin behaves differently from somatostatin in the experiment, it is a good idea to identify precisely how they differ, to find out what the actual amino acid sequence of llamastatin is. In other words, what do the new bases code for? To do this, you must first know which strand of DNA in the synthetic gene is the coding strand (Box 5.1). You were told that it is the lower strand (Figure 6.4). We can thus see that the two new DNA triplets, TAC and GAA, would be transcribed to give the complementary mRNA codons AUG and CUU, respectively. Reference to Table 5.1 shows these codons to code for Met and Leu, respectively. Therefore llamastatin must be identical to somatostatin except for Phe at position 6 and Cys at position 14 being substituted by Met and Leu, respectively (shaded pink in Figure 6.6a).

Looking at Figure 6.5, you can see that changing the Cys at position 14 to Leu means that, unlike somatostatin, llamastatin can have no disulphide bridge. But the other change, at position 6, is more significant in explaining this experiment. Consider what would happen if a hybrid polypeptide between β-galactosidase and llamastatin was treated with cyanogen bromide. The β-galactosidase would be fragmented and the Met before the first amino acid in llamastatin (the N-terminal Ala) would be affected, as intended, as done by Itakura and colleagues for somatostatin (Figure 6.5e). But the Met at position 6, *within the llamastatin proper*, would also be affected. So the llamastatin itself would be fragmented into two pieces (Figure 6.6b). So there would be no (intact) llamastatin. It is not surprising that an antibody specifically prepared against llamastatin might not bind to such fragments and hence the failure to detect any sign of llamastatin in the experiment—of course there isn't any. Likewise, the biological assay also comes up negative.

(Incidentally, say llamastatin did have a disulphide bridge between positions 3 and 14, as does somatostatin; then, even though cyanogen bromide splits the chain at position 6, the two halves would still be held together via the disulphide bridge. But as the other difference between llamastatin and somatostatin, a Leu at position 14, removes a

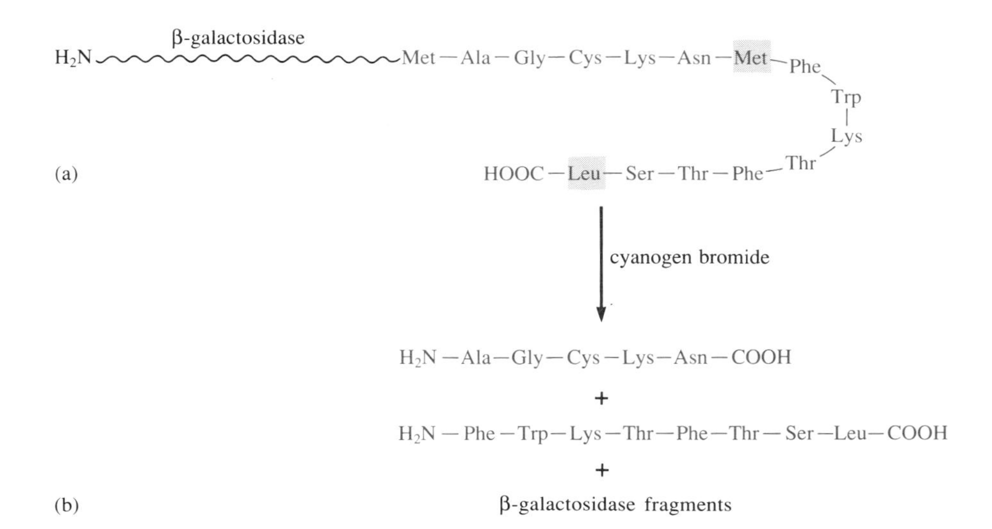

Figure 6.6 Cleavage of the hybrid β-galactosidase–llamastatin polypeptide by cyanogen bromide. Treatment of (a) the hybrid polypeptide with cyanogen bromide produces (b) fragments of β-galactosidase plus two fragments derived from llamastatin itself by cleavage at position 6.

Cys and hence any possibility of a disulphide bridge, no such holding together of the two llamastatin fragments occurs. They separate; Figure 6.6b.)

Of course, in one sense we have set this problem backwards. Presumably, in 'reality' (leaving aside the lack of reality of llamastatin), you would know the different amino acids were Met and Leu *before* you synthesized an appropriate synthetic gene. So you would know about the internal Met in somatostatin *in advance*. So in contrast to the elegant pre-planning of Itakura and colleagues to use the *additional* Met (before the N-terminal Ala) as a suitable point to release somatostatin from the hybrid polypeptide, using such a strategy for llamastatin would be plain dumb! Some other method of separating the llamastatin from the hybrid polypeptide would be needed and an additional Met before the N-terminal Ala would probably then be irrelevant.

Activity 7.1

(a) A cDNA. FVIII is synthesized initially as a polypeptide comprising 2351 amino acids. As the genetic code is triplet, there need be $3 \times 2351 = 7053$ bases in the coding strand of a gene for this polypeptide; that is 7053 base pairs in the gene (i.e. counting both strands of the DNA). Yet the natural gene for FVIII is actually much much longer — 186000 base pairs. The logical conclusion is that the gene for FVIII contains one or more introns (in fact it contains 25). Therefore it could not be expressed in *E. coli* as prokaryotes cannot carry out post-transcriptional modification of mRNA (Box 5.2). Using a cDNA derived from cytoplasmic mRNA for FVIII would be the correct alternative. (You may have also realized, that as the gene for FVIII is very long, it would be difficult to introduce it into *E. coli* as a restriction enzyme fragment. In fact it would be impossible. As mentioned in Section 7.1, even λ vectors can only accept up to around 24000 base pairs of foreign DNA. Though other vectors with greater capacity do exist for *E. coli*, even these fall way short of 186000 base pairs.)

(b) To get a cDNA for FVIII we must chose cells that contain mRNA coding for the polypeptide. As the natural site of FVIII synthesis seems to be the liver, liver cells would do. In fact, only a small amount of mRNA for FVIII is found there. Nevertheless, liver cells can provide a source of mRNA suitable for producing cDNA for FVIII.

(c) No. In humans the initial polypeptide produced from translating the mRNA for FVIII undergoes 'considerable post-translational modification'. It is unlikely that *E. coli* can carry out all such modifications and a host cell closer in species origin to human FVIII would be appropriate (Section 7.2.2). Some mammalian tissue culture cells might be suitable. In fact a monkey cell culture has proven so. You might also have mentioned the potential additional purification problems using an *E. coli*-derived product for human therapeutic use; i.e. to ensure total absence of the cell wall toxin (Section 7.2.1).

(d) Cost and availability might be among the reasons — in principle at least, a source such as from genetically engineered tissue culture cells *might* ultimately prove easier than isolating it from donated blood. However, at the time of writing (autumn 1992), no such source has yet reached that stage of development. But another reason is safety. Given the well-publicized dangers of blood-borne viruses such as HIV and hepatitis, a source of FVIII other than human blood might well be desirable.

Activity 8.1

As always, descriptive questions do not elicit hard-and-fast answers, but your answers should be along the following lines:

Part I

(a) Twin motivations: (i) *humanitarian*—medical research aimed at curing cancer in humans; (ii) *commercial*—OncoMouse sold by a commercial organization, DuPont.

(b) To ensure proprietary rights on OncoMouse. Patent ensures exclusive production rights or royalties from other producers licensed by the patent holders. (In fact, the patent holders on the OncoMouse are Leder and Stewart of Harvard University, but DuPont sponsored the work and have marketing rights.)

(c) *Ethical/humanitarian*—(i) OncoMouse will suffer (it is *designed* to develop cancer); (ii) some implication too, *perhaps*, of distaste for deliberate 'designing' of animals for human exploitation.

(d) Names of groups, quoted statement and suggestion of fear of further patents imply that they are against what they regard as inhumane treatment of 'sentient being(s)', in general.

Part II

Some of the factors you might have considered are listed below. (*Note*: there is some overlap of the factors given here with those in Part I. There is also overlap between the factors in the list, for some of them are difficult to categorize clearly.)

1 The science itself—are the methods and knowledge available?

2 Motivation—is the development worthwhile? Curing or preventing disease, improving crop productivity, making money might be among motivations.

3 Technology of scale-up—can we scale up the process to a suitable industrial level?

4 Economics—is the project commercially viable? Does it suggest a new saleable product? Are there rival products or processes? How do costs compare? What time-scale? Is development finance available?

5 Safety—is the process safe, are products safe? Safety of workers and the consumers/patients, environmental safety (pollution, ecological effects)?

6 Political considerations—is the development seen as desirable in the 'national interest' (e.g. share of world market, technological development)? (See point 7, below, also.)

7 Public perception—is the development seen as desirable/undesirable, safe/unsafe? How does such perception, *whether well-founded or not*, affect the development; effect of mass media, lobby groups, political effects, etc? (See point 8, below, also.)

8 Ethical considerations—what is the public attitude? Fears of 'tinkering with nature', distaste over altering animals genes, 'prospects' for genetic engineering of humans?

9 Broader social/ethical consequences of developments (employment, effects on trade with Third World).

You may well have thought of other factors.

Activity 9.1

The story

Dr Craig Venter, working at the NIH, determined the base sequence of more than 2000 fragments of genes, which he obtained from existing libraries. The identity of the gene fragments, and the function of any proteins they may encode, is unknown.

The NIH applied for patents on the fragments. After some debate, the patent office refused the patents on the grounds that the sequences were not novel, nor were they useful (although they might become useful when and if their functions are discovered), and they were 'obvious', i.e. a direct consequence of previously published work, and therefore not patentable.

The NIH's position

If the NIH published details of the DNA sequences without first patenting them, then it would not be able to share in the profits of any future application of the information. Any future work would be obvious, and so unpatentable. This might deter companies from developing useful drugs based on the sequences. Although there is no guarantee that patents on these sequences will guarantee patents on later, fuller, descriptions (indeed, there is a strong possibility of the *opposite* happening), NIH is going to appeal against the patent office's decision. If NIH does not apply for patents, its employees might.

The biotechnology industry's position

This is divided. The Association of Biotechnology Companies (ABC) is in favour of patenting the sequences. But the Industrial Biotechnology Association (IBA) does not think that partial sequences of unknown function should be patented. There are also worries that licensing costs might make product development even more difficult.

Activity 9.2

On the positive side, the knowledge that John X may be made ill by substances in the working environment might lead to an effort on the part of the employers to clean up the workplace. They might offer him an alternative job where contact with the oil could be avoided. The employers might also take the 'healthy worker' approach, and encourage all their employees to be screened. Finally, if they were particularly altruistic, they might fund research into cures for the diseases caused by the damaging substances.

On the negative side, the knowledge that John X only *might* develop a disease caused by a workplace pollutant may be enough to let the employers off the hook as far as cleaning up the environment is concerned. Both John's insurers and those representing the employers might raise insurance premiums significantly if John remained in his present job. There would thus be considerable pressure on John to leave: this might result in a social stigmatization, and would almost certainly result in a lot of personal distress. If you doubt this, think of the pressure currently exerted on people who wish to smoke in public.

Activity 9.3

Three important factors that should be considered are outlined below.

1 Is the test safe? That is, is the sample needed for testing easy and safe to take from the individual? (e.g. a blood sample would be safe).

2 Is the test reliable? Does it give an accurate answer, rarely giving false results? A false diagnosis can obviously be very undesirable.

3 What are the ethical consequences of a test being available? How would individuals told that they were homozygous (and would therefore develop the disease) react to the knowledge? Would individuals be coerced into undergoing tests—if so, is this desirable? How would carriers (i.e. heterozygotes) respond to the knowledge? Would this enable them to make better, more informed, choices about having children, for example? Is permission also being sought to use the test *in utero*—or will such permission be the next step? What further ethical questions does this give rise to? Will the information about *lethal* be confidential—will insurers/employers/other family members try to obtain the information?

Activity 10.1

(a) The answer depends on your perception of the problem. The main worries raised by the release of a genetically engineered organism are:

1 The organism might compete with existing varieties so successfully that it leaves them no space or resources for growth. This is a problem of *species loss* (already an area of concern because of the increase in *monoculture*, i.e. the widespread cultivation of only one, or a few species, of each crop plant).

2 The engineered genes might 'escape' and be transferred to other species where they might have adverse effects. As these new gene combinations might never have existed before, there will be no natural barriers to their spread. They might eventually interfere with the natural carbon and mineral cycles upon which all our lives depend.

(b) Whether you believe that plants are inherently safer than microbes affects your answer to this question. There is a general public perception that 'bugs are dangerous', yet, scientifically, this is not substantiated. *Natural* mechanisms, distinct from normal reproduction (which involves individuals from the *same* species), exist for transferring genes *between* many species; it is not known how commonly these processes occur in higher organisms, but since they *can* occur, it is likely that they have already done so, with no ill-effects that we are aware of.

It should not be beyond the wit of genetic engineers, if they are required to do so, to incorporate some kind of fail-safe device into genetically engineered plants. This might, for example, enable the plant to survive only under stringent conditions. Even if an 'escape' occurred, no spread would be possible.

You might also have pointed out that, for all commercial products, built-in obsolescence ensures a continuing market for the product. No manufacturer wants a light bulb that lasts forever, so why should anyone engineer a plant that might do so? A 'self-destruct' mechanism, that came into operation after a given time interval, might please both manufacturers and environmentalists.

No doubt you thought of other arguments.

Activity 10.2

First identify a *source* of the protein conferring the property of enhanced liquid absorption. We shall assume that it is the product of a single gene, the *absorb* gene.

Starting with the specific tissue producing the *absorb* protein, isolate the population of mRNAs produced here. Then make cDNAs of these mRNAs, clone the cDNAs and screen the clones (using the protocol in Figure 5.6). Since *absorb* mRNA will probably be a major mRNA component of the source tissue, the number of clones needed to be screened should not be too large (Section 5.3.2). (Alternatively, if the *absorb* protein is of known molecular mass, it *might* be possible to isolate *absorb* mRNA by size alone, make *absorb* cDNA and clone it.)

The *absorb* cDNA could then be inserted into an expression vector known to be active in plants, such as the modified T_i plasmid (Section 8.3). (You would want to use an expression vector, because we need the protein product of the *absorb* 'gene' to be produced.) It would be important to put the *absorb* gene under the control of a tissue-specific regulatory sequence—in this case, one which was active only in cotton fibres. This could probably be found close to a fibre-specific gene. The expression vector would then be inserted into cotton cells, and whole cotton plants regenerated, probably from callus culture. This might take some time. When fibres were produced, the *absorb* gene would be expressed, and the cotton fibres would have enhanced water absorption.

Activity 10.3

(a) Read the following suggested article and compare it with your own in terms of ease of reading and information content. Yours may well be better; try to identify what you feel are the weak and strong points of what you have written.

The use of genetically engineered BST is controversial, both in the UK and the USA. The scientific consensus is that BST, able to boost milk yields from cows by up to 25%, is safe; its natural form is present in all milk, tests show it is non-toxic and when ingested its breakdown renders it biologically inactive in humans. In the USA, successive endorsements of BST by the Food and Drug Administration and independent experts have failed to reassure sceptics or prevent boycotts of dairy products. Consumers are sceptical of the new technology that BST represents. Producers fear that BST use would lower milk prices and threaten livelihoods. In Europe, the flexing of political muscle between the regulatory experts that comprise the Committee for Veterinary Products and the Directorate General VI, with overall responsibility for agriculture, has contributed to delay and confusion. The DG VI is concerned with the 'social impact' of BST, including its influence on the CAP; the CVMP with quality, efficacy and safety. Opinion polls of dubious objectivity, as well as political in-fighting, have helped tarnish the image of effective and centralized EC regulation. Now DG VI has proposed a further moratorium on the use of BST for two years, a reflection more of a desire to clip the wings of the CVMP than of sound scientific thinking.

(b) (i) The case *for* administration of BST to dairy cows:

1 BST can increase milk yields by up to 25%.

2 Because milk will become more plentiful, it will cost less.

3 rBST is no different from normal BST.

4 The milk produced by treated cows is completely safe for human consumption. The levels of BST found in milk from treated cows are at the upper limit of those found in untreated cows.

(ii) The case *against* administration of BST to dairy cows:

1 Regular injections of BST, while increasing milk production, also increase the incidence of mastitis.

2 BST-treated cows suffer reproductive failure.

3 The dairy industry is already so intensive that cows have seriously impaired immune systems and so succumb to serious infections.

4 Because milk prices will drop if BST is widely used, small farmers may be forced out of business.

5 We do not need more milk; there is already a milk lake in the EC.

6 There is evidence of consumer prejudice against milk from BST-stimulated cows.

Note: this activity relates to injected BST protein (a product of a genetically engineered bacterium), *not* to cows injected with the *gene* for BST. This latter prospect (Section 10.7) might still raise similar arguments to those considered above for injected BST (*protein*). You might like to think of where the arguments would be the same and where, possibly, different.

Acknowledgements

The Course Team would like to acknowledge the help and advice of the external assessor for this book, Dr Robert Old of the University of Warwick.

Grateful acknowledgement is also made to the following sources for permission to reproduce material in this book:

Extracts

Extract 8.1 Hawkes, N. (1991), 'Patent granted for cancer mouse', *The Times*, 15 October 1991, © Times Newspapers Ltd 1991; *Extract 9.1* 'Intellectual property Obviously not', *The Economist*, 3 October 1992, © The Economist, September 1992; *Extract 10.1* Miller, H. I. (1992), 'Putting the bST human-health controversy to rest', *Biotechnology*, February 1992, p. 147, Nature Publishing Co.; *Extract 10.2* Vandaele, W. (1992), 'bST & the EEC: Politics vs Science', *Biotechnology*, February 1992, pp. 148–149, Nature Publishing Co.; *Extract 10.3* Blythman, J. (1991), 'Will daisy become a monster', *The Independent,* 29 June 1991. Cartoon: Michael Heath/The Independent.

Figures

Figure 1.1 Courtesy of Mercedes Benz (UK) Ltd; *Figure 1.2* Reproduced by permission of *The Wall Street Journal*, © 1980 Dow Jones & Company Inc. All rights reserved worldwide; *Figure 1.3* Wilkie, T. (1991), 'Mice embryos' sex changed', *The Independent,* 9 May 1991. Photograph: courtesy of Dr Robin Lovell-Badge/Medical Research Council. Cartoon: Colin Wheeler. Hosking, P. and Counsell, G. (1991), 'Nationwide cuts home loan rate to 12.25%', *The Independent,* 9 May 1991; *Figure 2.1* Reproduced by courtesy of Dumbarton Oaks/Jill Leslie Furst; *Figure 2.2* Bingham, J. (1979), 'Wheat breeding objectives and prospects', *Agricultural Progress*, **54**, Agricultural Education Association; *Figure 2.5* Martin, G. L. (1991), 'Thorny problem of blue rose solved', *Daily Telegraph*, 15 August 1991, © The Telegraph plc, 1991; *Figure 3.1* University Museum, University of Pennsylvania; *Figure 3.2* Rijksmuseum van Oudheden; *Figure 3.3* Kyowa Hakko Kogyo Ltd, Tokyo, Japan; *Figure 4.1b* Stanley N. Cohen, Stanford University; *Figure 7.1* From Williams, D. C. *et al.* (1982), 'Cytoplasmic inclusion bodies in *Escherichia coli* producing biosynthetic human insulin proteins', *Science*, **215**, pp. 687–689; *Figure 7.3* Singer, M. and Berg, P. (1991), *Genes & Genomes*, Blackwell Scientific Publications Ltd; *Figure 8.1* Courtesy of Prof. E. C. D. Cocking, University of Nottingham; *Figure 8.3* From Schell, J. and Montagu, M. V. (1983), 'The T_i plasmids as natural and as practical gene vectors for plants', *Biotechnology*, April 1983; *Figure 8.5* From Old, R. W. and Primrose, S. B. (1989) *Principles of Gene Manipulation*, 4th edn, Blackwell Scientific Publications Ltd; *Figure 8.6* The front cover of *Science*, November 1983, **222**, No. 4625, © American Association for the Advancement of Science; *Figure 9.4* Carter, C. O. (1962), *Human Heredity*, Penguin Books, © C. O. Carter 1962; *Figure 9.8* Reprinted with special permission of King Features Syndicate; *Figure 10.1* The authors are grateful to the John Innes Institute for permission to publish this photo; *Figures 10.2 and 10.5* Dr G. M. Gadd; *Figure 10.3* Nature Publishing Co.; *Figure 10.6* Mahon, G.; *Figure 10.8b* From Willadsen, S. M. and Polge, C. (1981), 'Attempts to produce monozygotic quadruplets in cattle by blastomere separation', *The Veterinary Record*, **108**, pp. 211–213.

Tables

Table 9.1 Courtesy of the Office of Health Economics (1983); *Table 10.2* Norris, J. R. (1978), 'Microbial control of pest insects', in Bull, A. T. and Meadow, P. M. (eds), *Companion to Microbiology*, Longman Group UK.

Index

Note: Entries in **bold** are key terms. Page numbers in *italics* refer to figures and tables.